TRADE UNIONISM

TRADE UNIONISM
PURPOSES AND FORMS

ROSS M. MARTIN

CLARENDON PRESS · OXFORD
1989

Oxford University Press, Walton Street, Oxford OX2 6DP
Oxford New York Toronto
Delhi Bombay Calcutta Madras Karachi
Petaling Jaya Singapore Hong Kong Tokyo
Nairobi Dar es Salaam Cape Town
Melbourne Auckland
and associated companies in
Berlin Ibadan

Published in the United States
by Oxford University Press, New York

British Library Cataloguing in Publication Data
Martin, Ross M.
Trade unionism: purposes and forms
I. Trade unions
I. Title
331.88
ISBN 0–19–827710–5

Library of Congress Cataloging in Publication Data
Martin, Ross M. (Ross Murdoch), 1929–
Trade unionism: purposes and forms/Ross M. Martin.

Bibliography: Includes index.
ISBN 0–19–827710–5
1. Trade-unions. 2. Industrial relations. I. Title.
HD6483.M345 1989 331.88—dc19 88–34557

Set by Litho Link Limited

Printed in Great Britain
by Biddles Ltd.
Guildford and King's Lynn

To Elena

PREFACE

Each of the two parts into which this book is divided had its genesis in a methodological problem. In the case of Part I, 'The Purposes of Trade Unions', the problem was a conventional definition of 'trade union' that manifestly excluded many, possibly most, organizations in the world laying claim to the title. The fact that a reference to 'purpose' was the excluding element in the definition led to an investigation of differing ideas about the purposes appropriate to trade unions—and, eventually, to a classification of them; as well as a different definition of 'trade union'.

Part II, 'The Forms of Trade Unionism', originated in a search for a sensible typology of national trade union movements. Once formulated and substantiated, the typology itself prompted an exploration of the reasons for the variations it delineated—and, eventually, a theory explaining them.

The distinction between the two parts may be stated, more crisply, in another way. Part I is concerned with the rhetoric surrounding trade unionism as an object of desire. Part II is concerned with the actual, national circumstances of trade unionism as an object of study.

Thus while both parts focus on trade unionism, they otherwise deal with quite disparate matters. One consequence of this is that they each rely on an almost totally different range of sources. They share, in fact, little more than a dozen references. This explains the separate bibliographies at the end of the book. Moreover, these are not merely separate, but differently organized. The one relating to Part I is a conventional, full list of all sources consulted, whether or not cited. The bibliography relating to Part II (as explained more fully in its own introduction) differs in two ways. In the first place, it is limited to sources either actually cited or, otherwise, of quite specific value to the work. Secondly, it is divided into categories which, while marginally complicating the task of those dedicated to the ultimate pursuit of footnote references, greatly simplify access for readers with broader interests.

This book owes much to others. I had the great advantage of comments on drafts of it from William Brown, Hugh Clegg, Leon Glezer, and Roderick Martin, all of whom read both parts; from

Richard Hyman and Michael James who read Part I; and from Malcolm Rimmer who read Part II. I also had help, mainly in connection with Part II, with sources and/or specific items of information. The assistance of Bruce Jacobs, Robin Jeffrey, and John Miller was especially useful. Others to be acknowledged in this respect include Dennis Altman, Ferenc Feher, Michael Forde, Kevin Hince, Michael James, Mave McDonagh, Angus McIntyre, Richard Mitchell, Christopher Neale, Liam O'Malley, Elliott Perlman, Rudolf Plehwe, Malcolm Rimmer, Hélène Teichmann, Ken Thomas, Claudio Veliz, and Robert Zuzovsky—together with the Canadian New Democratic Party, the Fiji Labour Party, the New Zealand Council of Trade Unions, and the Swedish Trade Union Confederation. Two of my undergraduate students, Cameron Lachlan (1987) and David Wales (1985), challenged their lecturer and in each case led me to change my mind on a point of interpretation relating to Part I. Annemarie Flanders devised the index with characteristic care and skill; and gave valued advice. The staff of the Borchardt Library, La Trobe University, Melbourne, worked wonders in the cause of assuaging the book's voracious appetite for references.

Marilu Espacio, as in all our other joint enterprises, once again reduced my flood of ugly, arrowed drafts to order with infinite forbearance and unfailing efficiency. The administrative genius of Talis Polis created a departmental haven that optimized my writing opportunities. A conversation with Leon Glezer, in 1981, first nudged me decisively in the direction of this book. The most profoundly personal debt, specific to the writing of it, is acknowledged in the dedication.

R.M.M.

August 1988

CONTENTS

PART II. THE FORMS OF TRADE UNIONISM

TABLES

PART I

THE PURPOSES OF TRADE UNIONS

A STUDY IN MODERN POLITICAL IDEAS

INTRODUCTION

What are trade unions for? The question, in one form or another, is often asked. Occasionally, it is posed as the prelude to a discourse on the inherent evils of trade unionism; and sometimes as the first step in an argument about the declining need for trade unions in modern society. More often, however, it paves the way for a statement of the 'actual', 'real', or 'true' purpose of trade unions. Often, too, that statement has been taken, if only implicitly, as a necessary part of the definition of a 'proper' or 'genuine' trade union. The outstanding feature of such statements of trade union purpose is their great variety. This part is about the nature of those variations.

John Dunlop once pointed out that 'theories of the labor movement' had sought to explain four key issues relating, respectively, to the origins of trade unionism, the growth of trade unions, 'the ultimate goals of the labor movement', and the reasons why workers joined trade unions.[1] Dunlop then went on to outline the ideas of eight writers on these issues. In doing so, he approached their ideas as competing descriptions and explanations of empirical reality.

The present work covers much the same writers—and many more. It is, however, concerned only with that aspect of their theories which relates to what Dunlop described as the 'ultimate goals', the purposes, of trade unions. In addition to this narrower focus, there is also a radical difference in approach. I am not concerned with assessing the accuracy of various theories of purpose as descriptions or explanations of empirical reality. I am concerned with them simply as *ideas*. I approach them, in other words, as essentially normative or ideological statements—that is, propositions about the purposes that trade unions *ought* to pursue.

Nor is there, in the following pages, any intention of evaluating the various theories in normative terms. That task is left to readers to perform in the light of their own ideological preferences. My aim has been the more modest one of surveying, as straightforwardly as possible, the remarkable variety and nuances of the

[1] In Lester and Shister, *Insights into Labor Issues*, pp. 164–5.

purposes which have been wished on the trade union during its relatively short history as an institutional form.

I have taken my illustrating authors as I have found them. My attention has been directed to what they have had to say on the single issue of trade union purpose. My categories, as a result, focus on similarities that have to do with that theme, regardless of disparities on others. The outcome, in conventional political terms, is some strange bedfellows.

The ideas and the circumstances of the different authors are presented below in widely varying degrees of detail. So far as ideas are concerned, a simple shortage of information acted as a constraint in a great many cases. (Marx, for example, would certainly have required more extensive treatment in Chapter 4 had he written a single, studied account of his views on trade unionism under capitalism comparable to Lenin's *What is to be Done?*) Otherwise, the attention paid to an author's ideas was primarily the product of three considerations: the ideas' pertinence as illustrations of a broad line of thought, or its development; their relative complexity or simplicity, as expressed; and, finally, their treatment in the secondary literature—which often amounted to total neglect.

So far as circumstances are concerned, these were dwelt on wherever (and to the extent that) it seemed they could aid an understanding of related ideas, or their development. This consideration partly explains the exceptional detail of the account, given in Chapter 6, of Lenin's views on trade union purpose under socialism. The extended nature of that account, however, also owes much to three other considerations—the literature's inadequate treatment of the shifts in Lenin's position; the subtleties in his own statements of it; and, finally, my hunch that this account will evoke hotter opposition than any other interpretation offered in Part I (which suggested the need for closer argument and more meticulous substantiation than would otherwise be required).

Three final points need to be made. The first involves the use of 'purpose' and 'function'. In much of the literature, these terms are used interchangeably. I propose to stand by the dictionary (*OED*) definitions of the two words because they involve a useful distinction between, on the one hand, an aim, goal, or objective (purpose) and, on the other hand, a method, means, or mode of action (function) intended to fulfil that purpose.

Secondly, there is a complication to do with the meaning that 'purpose' bears in this book. Mostly, as the preceding and following paragraphs illustrate, the term is adequately interpreted in its commonplace meaning—as denoting a goal in the sense of the *thing* to be achieved (for example, higher wages). Occasionally, however, it is used in a context which places the emphasis on a quite different aspect or dimension of purpose. This relates to the putative *beneficiaries* of a goal's achievement (for example, the working class). The distinction between the two dimensions is elaborated below, on page ten.

The third point also concerns 'purpose', but only in the sense of the goal (thing) to be achieved. In this respect, there are what may be described as layers, or levels, of purpose. Some purposes attributed to trade unions are distinguishable from one another as alternative purposes. Others are distinguishable in terms of primary and secondary, major and minor, purposes. But others again, while notionally distinguishable, cannot sensibly be either treated as alternatives or ranked in order of priority. This occurs in the case of specifically formulated purposes which clearly complement each other in the sense that one implies, embodies, or subsumes the other. For example, the familiar purpose of improving wages and working conditions implies the purpose of industrial self-government in at least some measure—or, working the other way, the latter purpose may be regarded as embodying the former. In either case, it seems reasonable to view them less as distinctive purposes than as different layers of the one purpose.

I

THE PROBLEM OF PURPOSE

The purpose of an institution is an obvious point of purchase for anyone wishing to understand it. In Mancur Olson's words: 'The logical place to begin any systematic study of organizations is with their purpose.'[1] Purpose, more often referred to as 'goal', has in fact figured prominently in the formal definitions of organization theory.[2] And quite apart from definitions, few contributions to this area of theory, as Herbert Simon has pointed out, 'manage to get along without introducing some concept of "organization goal" '.[3]

At the same time, organization theorists have become highly sensitive to the empirical difficulties of isolating purpose in the case of specific organizations, and especially the kind of purposes which Simon termed 'over-all goals'.[4] They have paid particular attention in this connection to the displacement of goals, the distinction between formal and 'real' goals, and to variations in the goals of the individuals and groups comprising an organization. Some, like Simon, have nevertheless maintained that the notion of purpose still has its uses at the empirical level.[5] Others, like Amitai Etzioni, have concluded that its inadequacies in application make an alternative approach necessary, but have still continued to lean on the notion in their formal definition of 'organization'.[6]

Writers wrestling with the problem of purpose in the context of organization theory have tended to skirt around trade unions. Their examples, for the most part, have been drawn from other types of organizations. But they have not sought to exempt the trade union from their generalizing propositions.

[1] Olson, *The Logic of Collective Action*, p. 5.

[2] See Barnard, *The Functions of the Executive*, p. 82; Philip Selznick in Rose, *The Study of Society*, p. 231; Talcott Parsons in Etzioni, *Reader on Complex Organizations*, p. 33; Etzioni, ibid. 3.

[3] In Etzioni, *Reader on Complex Organizations*, p. 158.

[4] Ibid. 174.

[5] Ibid. 158–74.

[6] See Etzioni, *Comparative Analysis of Complex Organizations*, pp. xi n., 132–6.

TRADE UNIONS DEFINED

The classic definition of a trade union was formulated by Sidney and Beatrice Webb in their pioneering study, *The History of Trade Unionism*. It occupied, appropriately, the first sentence of their first chapter.

> A trade union, as we understand the term, is a continuous association of wage-earners for the purpose of maintaining or improving the conditions of their employment.[7]

Although their book dealt almost exclusively with Britain, the Webbs thought of this definition as generally applicable. They made that clear twenty-six years later, in the second edition, when they deleted 'employment' and substituted 'working lives'. The change, they explained, was designed to obviate the unintended implication of the original that trade unions 'have always contemplated a perpetual continuance of the capitalist or wage system'.[8]

In both versions, the definition attributes to trade unions two distinctive features. One has to do with the nature of their *members* ('wage-earners'); the other with the nature of their *purpose* ('maintaining or improving the conditions of their [members'] working lives').

The first of these features can be regarded as acceptable and essentially uncontroversial as a defining characteristic—although not without some qualification. Describing the members of trade unions simply as 'wage-earners' is too restrictive in the realities of today's world, and was arguably so even in the Webbs' world. For it excludes 'salary-earners', nowadays more commonly described as white-collar or non-manual workers. The broader term 'employees' covers both manual and non-manual categories. There is no distinction to be made in this respect between private and public employment; and it is plainly inappropriate to assume (as the Webbs once did) that the status of employee is unique to capitalist systems.[9] So, as a term designating those who form the memberships of trade unions, 'employees' encompasses the assembly line operative in Detroit, the post office clerk in Bombay, and the factory director in Leningrad.

Thus a trade union may be defined as being *at least* an

[7] *History of Trade Unionism* (1894), p. 1.
[8] Ibid. (1920), p. 1 n.
[9] See ibid.

association with a membership confined to employees. Typically, too, it may be thought of as enclosing employees who, in S. C. Sufrin's phrase, share some 'commonality of employment'[10]—such as the same occupation or craft, the same employer or industry, the same group of industries or occupations, or simply the same locality.

Inevitably, there will be anomalies, or apparent anomalies. A few acknowledged trade unions in some capitalist societies accept as members contract or sub-contract workers (for example, owner-drivers in the road transport industry) who are not technically employees. But such exceptions, however striking in principle, are insignificant in practice. Again, there are associations with names like sports club, credit union, and social club which often confine their membership to employees with a particular commonality of employment. But, whether or not they are associated with a trade union (as is often the case), such associations are geared only to those among the relevant employee population who are interested in the specific activity of playing football, saving money, or wining and dining. The trade union, on the other hand, is geared to a category of employees *as such*. That is the basis on which the membership criterion works as a defining characteristic, irrespective of differences in geography, culture, and politico-economic system.

The case is quite different, however, when it comes to the second feature of the Webbs' definition—the nature of trade union purpose. The problem about purpose as a defining characteristic is that, in contrast to the membership factor, the nature of trade union purpose is neither self-evident nor uncontroversial.

VARYING PURPOSES

The major purpose of some organizations can be stated in a way which readily wins general acceptance. Thus Simon, using three examples, suggested that one might reasonably regard 'conservation of forest resources as a principal goal of the US Forest Service, or reducing fire losses as a principal goal of a city fire department', or (perhaps a little more arguably) 'profit as a principal goal' of business firms.[11] There is, however, no possibility of securing

[10] Sufrin, *Unions in Emerging Societies*, p. 8.
[11] Simon in Etzioni, *Reader on Complex Organizations*, p. 174.

anything like a similar degree of agreement when it comes to the major purpose of trade unions.

This purpose, as the Webbs stated it ('maintaining or improving the conditions of their [members'] working lives'), embodies a proposition about the satisfaction of interests. As such, it has two aspects or dimensions. One relates to the *possessors* of the interests to be satisfied (in the particular case, the members of the trade union). This goes to the issue of a trade union's *responsibility*. The other relates to the *nature* of the interests (in the particular case, working conditions). This goes to the issue of a trade union's *goals*. These two dimensions may be summarily delineated, in turn, by way of two questions. One is: 'Whose interests?' The other is: 'Interests in what?'

Conceptions of trade union purpose, similar to the Webbs', have commonly figured in definitions offered by others. Thus Peter Blau and Richard Scott, among the few organization theorists to have specifically grasped the nettle of the trade union, tossed it into a category of 'mutual-benefit associations' (as distinct from 'business concerns', 'service organizations', and 'commonweal organizations'). Mutual-benefit associations are characterized by the fact that they are 'expected to serve the interests of the rank and file' membership.[12] So much for the first dimension of purpose. As for the second dimension, Blau and Scott distinguished trade unions from other kinds of mutual-benefit associations by the nature of their 'specific objectives', which involved 'improving employment conditions'.[13] Olson endorsed these terms when he grouped trade unions among those organizations that are '*expected* to further the interests of their members'.[14] Again, the first dimension of purpose, followed by the second: 'Labor unions are expected to strive for higher wages and better working conditions for their members.'[15]

Among writers with a particular interest in trade unions, the use of similar conceptions of purpose as a defining characteristic has a long history which pre-dates the Webbs. Thus George Howell's depiction of trade unions as 'voluntary associations of workmen for mutual protection and assistance in securing generally the most favourable conditions of labour'.[16] And the popularity of this

[12] Blau and Scott, *Formal Organizations*, p. 44.
[13] Ibid. 252.
[14] Olson, *The Logic of Collective Action*, p. 6.
[15] Ibid.
[16] Howell, *Conflicts of Capital and Labour*, p. 128.

approach has remained high ever since the Webbs took it up. Moreover, on the part of English-speaking writers at least, there has been a pronounced tendency to regard this vision of purpose as both obvious and generally acceptable. Thus V. L. Allen:

> There is no difficulty about defining the aims of trade unions. It is acknowledged universally that their prime aim, and the one from which others flow, is to protect and improve the living standards of their members.[17]

But the comment is misplaced (even if Allen meant to confine it to capitalist economies or to Britain alone). The issue has in fact always been hotly contested. Tom Clarke recognized this in a reference to the 'considerable conflict since the origin of trade unions as to what their basic purpose was'.[18] On the other hand, he overlooked the complexity of the conflict when he reduced it to a simple two-way clash between promoting 'the sectional economic interests of particular occupational groups' or furthering 'the general interests of the working class in combatting capitalism'.[19]

TRADE UNION PURPOSE: THE IDEOLOGICAL RANGE

A number of theories of trade union purpose differ significantly from that embodied in Webbsian-style definitions. Each of them may qualify as explanatory or empirical in so far as they purport to delineate the attitudes and policies of specific trade unions, their leaders or members. But they take on a different character when employed (as they more commonly are) in a prescriptive vein to impute a 'proper' or 'true' purpose to the trade union as an institution, with the implication that all other attributed purposes must be either illegitimate or secondary. In this case, they are ideological propositions since their imputed validity depends not on empirical observation but on preferred values.

The ideological character of conceptions of trade union purpose is something which is rarely acknowledged in explicit terms. Richard Hyman and Bob Fryer are unusual in this respect (although not in their preference for 'functions' as a synonym for purpose): 'Evidently . . . debates about trade union functions are typically

[17] Allen, *Militant Trade Unionism*, p. 149.
[18] Clarke and Clements, *Trade Unions Under Capitalism*, p. 11.
[19] Ibid.

mediated by . . . ideological frames of reference.'[20] But, in the same paper, they also demonstrated the ease of the slide from analysis to prescription when they implied that *the* definitive trade union objective (though 'often . . . eroded in . . . actual practice') was 'workers' control'—a remark that followed disconnected but correspondingly dismissive references to a trade union purpose dependent on 'some hypothetical "public interest" ', and to 'the spurious trade unionism created by totalitarian political regimes'.[21]

As the comments of Hyman and Fryer insinuate, many purposes have been foisted on trade unions. Usually, they have been advanced in frankly prescriptive terms. Occasionally, they have been based on some form of empirical investigation and (to the extent that this is so) can be accepted as analytical or descriptive, although not necessarily as true, propositions. But even in these cases there is evident a tendency to assume, if only implicitly, that the discovered purpose is the *proper* purpose. On the other hand, it must also be said that all of the main ideological positions on union purpose are empirically relevant in the sense that there are cases (contemporary or historical) of their being officially espoused, in some form, by trade union bodies.

The various theories of trade union purpose are explored below in terms of five principal categories: pluralist, syndicalist, Marxist-Leninist, organicist, and authoritarian. These are broad categories, in that most of them embody quite striking internal variations. The leading characteristic of each category is signalled in the sub-title of the relevant chapter.

The sequence in which they are considered is not altogether arbitrary. In the first place, the five categories fall into two clearly distinct groupings according to whether the emphasis is on a trade union role which is essentially conflictual (pluralist, syndicalist, Marxist-Leninist) or essentially consensual (organicist, authoritarian). In the second place, the sequence comes close to constituting a spectrum on the issue of the scope of the responsibility imputed to trade unions: the pluralist conception is the narrowest, the organicist and the authoritarian the widest, while the syndicalist and the Marxist-Leninist represent an intermediate position in this respect. In the third place, the principal rationale for locating the syndicalist category alongside the pluralist is that they share an

[20] Hyman and Fryer, 'Trade Unions', p. 174.
[21] Ibid. 158, 174, 181.

assertion of trade union autonomy in relation to both party and state, whereas the Marxist-Leninist involves an assertion of autonomy in relation to the state alone. The autonomy issue also explains the positioning of the organicist and authoritarian categories in relation to each other: the former predicates an area of independent trade union action which the latter denies absolutely.

One further point needs to be made. It is that the various theories are each analysed in terms of the two dimensions of purpose specified in the preceding section of this chapter. These two dimensions refer, respectively, to the responsibility and to the goals attributed to trade unions. And, as we have seen, they are summarized in terms of two questions: (1) 'Whose interests?'; (2) 'Interests in what?'

2

THE PLURALISTS

Trade Unions as Industrial Regulators

Pluralists, like Marxists, regard conflict as the central feature of societies with market economies and liberal democratic political systems. But there the agreement ends. For while the Marxist identifies the crucial source of conflict as social classes, the pluralist puts it down to sectional interests that cut through and across classes—and, therefore (unlike the Marxist), attributes fundamental significance to the open competition among organized groups, especially those concerned with economic interests. Again, while the Marxist predicates that the conflict is ultimately destructive of the existing socio-economic system, the pluralist perceives it as basically supportive of the system, and even invigorating, because he believes that the intensity of inter-group competition is tempered by a broad acceptance of certain 'rules of the game'. Pre-eminent among these is the principle that neither the goals actively pursued by groups nor the methods employed to achieve them should threaten the fundamentals of existing institutional relationships. Otherwise, groups seek to realize their goals (according to this view) by bargaining with each other and/or with government; and the resultant agreements typically express compromises, rather than outright wins and losses.

Pluralist theories of trade union purpose are the narrowest and, on the face of it, the most modest of all such theories. It is the pluralist approach to the issue of purpose with which the Webbs' definition of a trade union has the closest affinity, although its classic statement occurs in Selig Perlman's *A Theory of the Labor Movement*, published in 1928. In its essentials, moreover, this approach is compatible both with the severe liberalism of free market economists, like Milton Friedman, and with the softer touch of some strains of non-Marxist socialism, as exemplified in the work of Allan Flanders.

WHOSE INTERESTS?

The answer to this question, concerning the first dimension of trade union purpose, was simply stated by Friedman. Apart from specific obligations imposed on them by the law, neither union nor business leaders ought to be regarded as having a responsibility beyond the interests of their shareholders or their members: 'the "social responsibility" of labor leaders is to serve the interests of the members of their unions'.[1] Flanders reiterated the point: 'The first and over-riding responsibility of all trade unions is to the welfare of their own members. That is their primary commitment; not to a firm, not to an industry, not to the nation.'[2]

This limited responsibility, for Friedman, was a crucial condition of a 'free economy'.[3] For Flanders, it was more broadly 'an essential part of the democratic process';[4] and, in contrast also, he displayed some concern for a wider, if secondary, social responsibility.

> Obviously trade unions cannot reasonably behave as if they were not part of a larger society or ignore the effects of their policies on the national economy and the general public. No voluntary organisations can do that with impunity. If they do, they turn society against them, and society can retaliate. In any case members of trade unions are citizens and consumers as well as producers.[5]

But he hastened to add: 'Even so, trade unions exist to promote sectional interests.'[6] And he emphasized that the interests concerned were members' interests 'as they see them, not their alleged "true" or "best" interests as defined by others'.[7]

The issue was not explicitly addressed by Selig Perlman, writing much earlier. However, his views on the second dimension of trade union purpose leave no doubt that, had he done so, he would have stated the responsibility of trade unions and their leaders in virtually identical terms.

INTERESTS IN WHAT?

There is rather more variety to pluralist versions of the second dimension of trade union purpose (the nature of the goals). But

[1] Friedman, *Capitalism and Freedom*, p. 133.

[2] Flanders, *Management and Unions*, p. 40; see also Flanders, *The Fawley Productivity Agreements*, pp. 235–6.

[3] *Capitalism and Freedom*, p. 133.

[4] *Management and Unions*, p. 41.

[5] Ibid. 40–1.

[6] Ibid. 41.

[7] Ibid. 40.

all of them focus on work-related matters of specific and intimate concern to union members themselves.

Selig Perlman

Too often, Perlman contended, the real goals of trade unions are obscured or distorted by outsiders (non-workers) who impute purposes which reflect their own ideological positions rather than the concerns of union members. He readily conceded that outsiders with an interest in trade unionism—'intellectuals' as he called them—had played an essential role in the development of a number of national labour movements. Nevertheless, he characterized them as 'labor's devoted but impractical friends'.[8] They performed a useful function so long as they confined themselves to research and advice on industrial safety, wage trends, and the like. But once these areas of specific and calculable expertise were abandoned, 'it is the rare intellectual who is able to withstand an onrush of overpowering social mysticism'.[9]

The problem, he thought, was that intellectuals' ambitions for trade unionism far exceeded those of union members whose different experience as manual workers gave rise to more modest ambitions expressed in their own ' "home-grown" ideology' ('the philosophy of organic labor').[10] In other words, intellectuals associated with labour, whatever the ideological differences among themselves, all shared a passion for committing trade unions to causes that were not as important to the members of those unions. The result was a continuing conflict within the labour movement between the ambitions of intellectuals and those of rank-and-file trade unionists. It was the influence of intellectuals that explained the occurrence of 'socialistic' unions.[11] But while the ambitions of the intellectuals were trumpeted abroad, those of trade unionists were less easily ascertained. They were, however, to be discerned in the actual customs and practices of unions. So, in order to distinguish 'fundamental from accidental purposes', to isolate 'the true purposes of unionism . . . from mere verbal pronunciamentos', one had to look at those customs and practices.[12] This was the claimed source of Perlman's theory of 'job-conscious unionism'.[13]

[8] Selig Perlman, *A Theory of the Labor Movement*, p. 300.
[9] Ibid. 281.
[10] Ibid. 6, 254.
[11] Ibid. 278.
[12] Ibid. 237, 278.
[13] Ibid. 274.

Trade unionism untouched by intellectuals, he concluded, was 'essentially pragmatic'.[14] Its pragmatism was a product of the manual worker's experience that the world he inhabited was 'a world of limited opportunity'.[15] For the worker was constantly confronted with a situation in which the number of those seeking work exceeded the number of jobs available. His outlook, as a result, was dominated by a 'consciousness of job scarcity'.[16]

It was this consciousness of scarcity which generated among workers an aspiration for security and, specifically, for 'job security'[17]—involving an assurance of both obtaining a job, when needed, and holding it once acquired. But security was not the only value at stake. There was also a type of liberty. Associated with the aspiration for security was the aspiration of workers for 'concrete freedom on the job'.[18] Concrete freedom involved the ability of a worker 'to face his boss "man to man"' without fear, a 'tangible sort of freedom' which 'matters supremely' to workers, as distinct from the kind of freedom that 'the intellectual talks about'.[19]

Job security and concrete freedom, in Perlman's schema, together constituted the *true* purpose of trade unions in the sense of the purpose which workers, left to themselves, wanted their unions to pursue. But, underlying security and freedom, there was something else. For both depended on the prior attainment of 'job control'.[20] Job control referred to a trade union's ability to control access to the 'job opportunities' available to its members.[21] Job control was seen as the one means by which the ultimate goals of job security and concrete freedom might be realized. Its achievement was an essential preliminary goal; and thus comprised a lower layer of trade union purpose.

Perlman distinguished two forms of job control. One, the ideal from the viewpoint of workers, he labelled 'union dictatorship'.[22] It occurred when the union, without otherwise impinging on the employer's role as 'owner of his business and risk taker', formulated and applied the rules governing its members' conditions of employment—and in this sense acted as 'the virtual owner and administrator of the jobs'.[23] Such unilateral regulation was rare, but the second form of job control was not. In this case, instead of establishing its 'ownership' of the jobs, the union established

[14] Ibid. 5. [15] Ibid. 239. [16] Ibid. 8. [17] Ibid. 9.
[18] Ibid. [19] Ibid. 290. [20] Ibid. 7.
[21] Ibid. 252. [22] Ibid. 263. [23] Ibid. 199.

'rights' in them through collective bargaining with employers, the rights being embodied in agreements that issued from the bargaining process. This kind of regulation Perlman called 'industrial democracy', 'a highly integrated democracy of unionized workers and of associated employer-managers, jointly conducting an industrial government with "laws" mandatory upon the individual'.[24]
individual'.[24]

It followed that the central *function* of the trade union, deriving from the purpose of job control, was the function of collective bargaining. There was as well, closely associated with this collective bargaining function, another function of great importance. It was an educational function. No trade union, Perlman remarked,

> can . . . be a union in the full sense of the word unless it has educated the members to put the integrity of the collective 'job-territory' above the security of their individual tenure.[25]

The odd thing about this is that, among the pluralist writers, Perlman stands alone in the explicit emphasis he placed on an educational function. In contrast, the idea that trade unions properly have the function of 'educating' their members (if in varying ways and to varying ends, as we shall see) is a matter of open and general acceptance in the case of all the other theories of union purpose.

The Economists

The inclination of classically capitalist economists to focus on economic factors has meant that they have also been inclined towards a narrow interpretation of trade union purpose. But, unlike Perlman, they have tended to emphasize wages. Thus Adam Smith's reference to 'combinations' of 'workmen', as being formed solely in relation to the wage issue,[26] was echoed two centuries later by F. A. Hayek who thought it undeniable 'that raising wages . . . is today the main aim of unions';[27] while Milton and Rose Friedman

[24] Selig Perlman, *A Theory of the Labour Movement*, p. 199.

[25] Ibid. 273; see also S. Perlman, 'The Principle of Collective Bargaining', where he talks of 'building discipline' (p. 154).

[26] Adam Smith, *The Wealth of Nations*, pp. 30–1.

[27] Hayek, *The Constitution of Liberty*, p. 275.

were only a little more circumspect when they wrote of the unions' 'major proclaimed objective—the wages of their members'.[28]

This kind of approach was elaborated in the 1940s by John Dunlop in a way that proved particularly influential among economists with a more specialized professional interest in industrial relations and labour economics. He equated business enterprises and trade unions on the argument that they shared the same primary purpose. That purpose was the maximization of the economic returns of their shareholders/members. The enterprise, in other words, sought to maximize profits; the union similarly sought to maximize 'the wage bill for the total membership'.[29] At the same time, however, Dunlop acknowledged that trade union wage policies might be directed at many objectives that had little to do with the wage bill.[30] The 'nonincome objectives of wage policy' could include such things as promoting the recruitment of union members and expanding employment opportunities.[31]

Arthur Ross, taking issue with Dunlop, expanded the qualifications. He argued against 'the mechanical application of any maximization principle'.[32] Trade union leaders and their members were neither as single-minded nor as economically sophisticated as Dunlop implied. Ross also saw serious flaws in the analogy between unions and business firms. Trade unions were more complex than firms in that they represented a greater diversity of interests, reflected in often sharp conflicts about union goals between such groupings as skilled and unskilled members, employed and unemployed, young and aged, women and men.[33] And they were more complex in another sense as well.

The formal purpose of the union [economic welfare] is vaguer than that of the business firm [profit]. Profit can be measured in one dimension. 'Economic welfare' is a congeries of discrete phenomena—wages, with a dollar dimension; hours of work, with a time dimension; and physical working conditions, economic security, protection against managerial abuse, and various rights of self-determination, with no dimension at all.[34]

[28] Friedman, *Free to Choose*, p. 242.
[29] Dunlop, *Wage Determination under Trade Unions*, p. 44.
[30] Ibid. 119.
[31] Ibid. 46–50.
[32] Ross, *Trade Union Wage Policy*, p. 8.
[33] See ibid. 31–4.
[34] Ibid. 27.

On this basis, Ross contended that a distinctively economic emphasis, however adequate in relation to business firms, was inappropriate in the case of trade unions because 'the trade union is a political agency operating in an economic environment'.[35] Its political character was, in part, a function of the nature of its 'formal purpose'—and that purpose, as Ross depicted it in the passage above, included Perlman's job security and concrete freedom in the shape of 'economic security' and 'protection against managerial abuse'. Moreover, in the course of a brief discussion of union 'loyalty', Ross also gave the barest hint of the educational function that Perlman had emphasized: 'If the union is effectively organized, the members will adopt its institutional needs as their own when mobilized for combat.'[36]

Allan Flanders

The significant development of Perlman's line of thought, however, occurred on the other side of the Atlantic. Flanders's essay, 'What are Trade Unions for?' was first published in 1968, precisely forty years after Perlman's book. The times were different. Full employment in advanced capitalist societies, unbelievable in the 1920s, was a reality: there was little purchase for a theory that explained union purpose principally as the outcome of a scarcity of jobs. From the end of the 1940s, too, a growing number of students of industrial relations had questioned the theory of job-conscious unionism on other grounds.[37] Flanders made no mention of 'scarcity', nor of Perlman. Nevertheless, while he put his own stamp firmly on the sequence and terms of the argument, the echoes of Perlman are unmistakeable in Flanders's discussion.

He started, like Perlman, explicitly from the assumption that the answer to the question of trade union purpose was to be found in 'the behaviour of trade unions'.[38] That behaviour, he argued, pointed unequivocally to collective bargaining as the activity, the function on which unions lavished most energy and attention. The reason why they did so was conventionally explained in terms of the Webbs' definition—in order to protect and promote their

[35] Ross, *Trade Union Wage Policy*, p. 12. [36] Ibid. 24.

[37] See, esp., Siegel, 'The Extended Meaning and Diminished Relevance of "Job Conscious" Unionism'.

[38] Flanders, *Management and Unions*, p. 41.

members' conditions of employment. But there was more to it than this. For collective bargaining was not only a means by which unions might raise wages and shorten hours. It was also 'a rule-making process', the rules being embodied in collective bargaining agreements.[39] Trade unions, that is to say, had an interest in *regulating*, as well as improving, conditions of employment. They sought the 'regulation or control' of employment relationships through collective bargaining.[40] And so to the affirmation of 'their basic purpose of job regulation';[41] or, again, their 'basic social purpose of . . . job regulation and control'.[42]

But Flanders, like Perlman, did not leave it at that. Trade unions did not have 'a bureaucratic interest in rules for their own sake',[43] so there had to be another layer of purpose. The point about the rules embodied in collective bargaining agreements was that they established rights and obligations. Above all, from the trade union perspective, they set out the recognized rights of union members in relation to their employers. What this meant was that, through the rule-making process of collective bargaining, the trade unions were able to construct 'a social order in industry embodied in a code of industrial rights' which lessened 'the dependence of employees on . . . the arbitrary will of management'.[44] It was a code, moreover, which protected not only trade unionists' 'material standards of living, but equally their security, status and self-respect; in short, their dignity as human beings'.[45] The echo of Perlman's notion of concrete freedom, as well as that of job security, is plain.

There was yet another layer of purpose. Collective bargaining also gave workers, through their trade unions, a more direct influence on the rules governing their employment than they would otherwise have had, even under state legislation. In short, it gave them 'participation in job regulation'; and such participation was the 'constant underlying social purpose of trade unionism' which was directed towards 'enabling workers to gain more control over their working lives'.[46] This, of course, was the essence and point of Perlman's notion of industrial democracy as well. And while such a goal may be modest enough by the standards of the bolder visions of trade union purpose explored below, it is certainly not without ambition. The role it predicates for the trade union is, in Hugh

[39] Ibid. [40] Ibid. [41] Ibid. 45. [42] Ibid. 46.
[43] Ibid. 41. [44] Ibid. 42. [45] Ibid. [46] Ibid.

Clegg's phrase, the role of 'industry's opposition—an opposition which can never become a government'.[47] It is a role that only pluralists claim for the trade union quite so absolutely.

[47] Clegg, *Industrial Democracy and Nationalization*, p. 22. This, it may be noted, echoes R. H. Tawney's earlier depiction of the trade union as 'an opposition that never becomes a government' (*The Acquisitive Society*, p. 151)—but Tawney was not ruling out the possibility of it becoming effectively a government: see Ch. 5 below. Clegg was: 'unions . . . must remain a permanent opposition'. *Industrial Democracy and Nationalization*, p. 33.

3

THE SYNDICALISTS

Trade Unions as Social Emancipators

Syndicalism had a brief heyday, mainly in the years before the First World War. As a plan of action, it was condemned by history to failure. Nevertheless, ideas associated with it have continued to echo in corners of some trade union movements.

As a school of thought, syndicalism consisted of three distinct strands. *Anarcho-syndicalism* was the earliest and, organizationally speaking, the most widespread and the longest-lived. It originated in France and was also particularly influential in Spain, Italy, Norway, and Latin America. *One Big Unionism* was American in origin, and had some influence in Canada, Great Britain, Australia, South Africa, and parts of Latin America. Although commonly described by its advocates as 'industrial unionism', this form of syndicalism was distinguished rather by an emphasis on organizational unity in a single comprehensive structure ('the One Big Union', it was often called). *Guild Socialism*, the youngest and most moderate of the three, originated in England; its influence appears to have been confined to Great Britain.

Three basic ideas, common to them all, linked these three strands of thought and identified them as syndicalist. One was the notion that the future socialist society would be characterized by what was usually, if loosely,[1] described as workers' control of industry. Such control, in contrast to the pluralist case, was not to be confined to the employment relationship. It was to include *all* aspects of an enterprise. Each industry would ultimately be administered by the workers employed within it, acting through their trade union. Syndicalists, in other words, envisaged the total dispossession of the private employer. So, of course, did 'collectivist' socialists like the Marxist-Leninists. The difference was that the collectivists wanted to replace the private employer with the state, whereas the syndicalists were anti-state in general and, in particular, rejected

[1] See Hyman, *Marxism and the Sociology of Trade Unionism*, p. 46 n.

state control of industry. For them, the capitalists, as controllers of industry, were to be replaced by their employees.

The second idea was that capitalism could not be overthrown by 'political' methods alone. The syndicalist point was that political parties were inherently incapable of achieving the true socialism of workers' control. The key to that lay, rather, in industrial organization and industrial methods—which meant the trade unions and their singular weapon, the strike. These were to be the decisive means of destroying capitalism.

The third common idea concerned the appropriate organizational principle for the trade unions through which workers were eventually to control industry. The principle was that of 'industrial unionism', requiring each industry to be covered by a single union which would recruit all employees within the industry regardless of their trade, occupation, or status. This was designed to facilitate both the achievement and the application of workers' control.

Underlying these shared ideas, however, there were considerable differences of emphasis and approach. On the issue of trade union organization, for example, the One Big Unionists favoured a high degree of centralization that was diametrically at odds with the Anarcho-syndicalist ideal. But it was the differences between Guild Socialism and the other two which were most marked and most varied. Indeed, the Guild Socialists did not consider themselves 'syndicalists'. Nor, for that matter, did those proud to call themselves syndicalists: they could denounce Guild Socialism as 'the latest lucubration of the middle-class mind . . . a "cool steal" of the leading ideas of Syndicalism and a deliberate perversion of them'.[2]

ANARCHO-SYNDICALISM

The fissures running through anarchist thought are many and deep. But there is one thing that all true variants of anarchism have in common, which identifies them as specifically anarchistic—and that is the goal of destroying the state, and replacing it with some form of ostensibly non-coerced co-operation between individuals and/or social groups. In addition, and of critical importance, the destruction of the state is to be accomplished with one blow, or not

[2] Quoted in Russell, *Roads to Freedom*, p. 91 n.

at all. The anarchist has no truck with the Marxist notion of an eventual, post-revolutionary withering away.

Anarcho-syndicalism differed from other variants of anarchism in its emphasis on trade unions. Where they breathed distrust of formal organization (preferring to emphasize individuals, social classes, or communal associations), Anarcho-syndicalism lauded the trade union. It was 'the organ of social transformation',[3] and 'superior to any other form of cohesion between individuals'.[4] Above all, it was the pure instrument of the working class, being directly controllable by that class. In contrast, socialist parties claiming to represent the workers were havens for middle-class intellectuals and bribable ex-workers. Through their involvement in the electoral and parliamentary apparatus of the capitalist state, they served capitalist interests by compromising the interests of the working class. Political action was not only futile; it was corrupting.

The working class had to act through the one institution that was peculiarly its own, the trade union. This, for many Anarcho-syndicalists, meant rejecting help from any non-working-class source (including, to the outrage of Errico Malatesta,[5] middle-class anarchists). It also meant relying on 'direct action'.

> Direct action is opposed to the indirect and legalised action of Democracy, of Parliament and of Parties. It means that instead of delegating to others the function of action . . . the working class is determined to act for itself.[6]

Direct action involved sectional strikes, boycotts, industrial sabotage, and, ultimately, the master-weapon of the working class acting through their unions—the general strike. This was to mark the climax of the class war. As a French trade union resolution of 1897 put it: 'the General Strike is synonymous with Revolution'.[7]

The Anarcho-syndicalist conception of the general strike is associated indelibly with the name of Georges Sorel, although he was not formally associated with the French syndicalist movement. 'It is in strikes', he claimed, 'that the proletariat asserts its existence.'[8] The general strike was the highest form of that

[3] Pierre Monatte in Woodcock, *The Anarchist Reader*, p. 216.
[4] Emile Pouget, quoted in Guérin, *Anarchism*, p. 80. [5] See ibid. 81.
[6] Hubert Lagardelle, quoted in Estey, *Revolutionary Syndicalism*, p. 74.
[7] Quoted in Spargo, *Syndicalism, Industrial Unionism and Socialism*, p. 91.
[8] Sorel, *Reflections on Violence*, p. 274.

assertion. But not *all* general strikes. A 'political general strike' might bring about apparent change, such as a constitutional reform; at best, however, it merely replaced one lot of non-proletarian rulers with another because it did not 'presuppose a class war'.[9] The type of general strike that would achieve real change was the 'syndicalist' or 'proletarian' general strike, which was a 'phenomenon of war' in that its aim was nothing less than the smashing of capitalism.[10] Only this kind of general strike was capable of achieving true socialism, 'proletarian Socialism'.[11]

Proletarian socialism meant two things above all. First, it meant a stateless society: 'Syndicalists do not propose to reform the State . . . they want to destroy it.'[12] Second, it meant direct working-class rule (and that required, among other things, the ousting from influence of 'sociologists', of 'fashionable people . . . in favour of social reforms', of 'intellectuals who have embraced the profession of thinking for the proletariat', and of all politicians, 'in whom the hunt for fat jobs develops the cunning of Apaches').[13] Again, as in the case of the proletarian general strike, the trade union was the key to proletarian socialism. For the trade union was more than just 'an organization of struggle', as Amédée Dunois remarked, it was 'the living germ of future society, and future society will be what we have made of the trade union'.[14] The same point was made less lyrically in the syndicalist programme adopted in 1906 by the French trade union confederation: the trade union of the future, it predicted, would be 'the group responsible for production and distribution, the foundation of the social organization'.[15] As the foundation of the future social—as well as industrial—organization, self-governing trade union bodies, loosely linked and freely co-operating with each other, would provide all the co-ordination required by a society which had no place for police, army, or civil service. The Anarcho-syndicalists, in fact, did not have a great deal more than this to say about proletarian socialism.[16] They were much more concerned with the close-at-hand drama of the struggle to achieve it than with working out the detail of what it would involve.

[9] Sorel, *Reflections on Violence*, p. 158.
[10] Ibid. 155, 159, 274. [11] Ibid. 157. [12] Ibid. 116.
[13] Ibid. 138, 151.
[14] Quoted in Joll, *The Anarchists*, p. 187.
[15] Quoted ibid. 186.
[16] See Ridley, *Revolutionary Syndicalism in France*, pp. 165–9.

ONE BIG UNIONISM

The One Big Unionists paid even less attention than the Anarcho-syndicalists to the details of the future society.[17] They were part of a movement that, for a time, put life into the idea of uniting all workers in one body—and not merely nationally, a dream almost as old as trade unionism itself,[18] but internationally. The organization they threw up had that ambition plain in its name: the Industrial Workers of the World. The IWW's pressing need was for members. It found them principally among unskilled and mainly immigrant workers in the United States. The organizing was hard, and conducted in the face of often violent opposition from employers, public authorities, and the conservative craft unions associated with the American Federation of Labor. The emphasis, accordingly, was on action rather than theory; on slogans rather than extended argument. Ideas, nevertheless, were still important to the men and women who came together in Chicago to form the IWW in 1905. They agreed that unionism should be organized along industrial rather than craft lines. But they disagreed on other issues; and their differences, especially on the matter of political action, three years later split them into two groups, both claiming the IWW label.

The majority faction, which retained the Chicago headquarters, was headed by William Haywood, a tough union official from the Western Federation of Miners. Under Haywood the Chicago IWW was committed to an unequivocally syndicalist position in terms of ultimate aim ('By organizing industrially we are forming the structure of the new society within the shell of the old'),[19] of industrial unionism, and of an ostensibly exclusive reliance on industrial action.

The minority faction settled in Detroit. Daniel De Leon, a former university teacher in law, was its dominant figure and simultaneously head of the tiny Socialist Labor party. It was De Leon who gave One Big Unionism much of such literary clarity and intellectual coherence as it could boast. Moreover, his inclination towards political action was based on an unusual interpretation of

[17] As illustrated in William Haywood's vagueness when cross-examined on the issue: see in Larson and Nissen, *Theories of the Labor Movement*, pp. 86–90.

[18] See Webb, *History of Trade Unionism*, pp. 113–6.

[19] Quoted in Brissenden, *The I.W.W.*, p. 352.

its role which, arguably, still left him firmly in the syndicalist camp.

De Leon was clear that the state was to be destroyed: 'in Socialist society the political State is a thing of the past'.[20] Its place would be taken by the 'administrative organs of the nation's industrial forces',[21] industry-based trade unionism: 'Where the General Executive Board of the [IWW] will sit there will be the nation's capital'.[22] And the transition to the new society depended upon the general strike—'the general lockout of the capitalist class', as De Leon preferred to think of it.[23] But, unlike the French Anarcho-syndicalists, De Leon believed that conventional political action had a part to play as well.

De Leon has been described as placing 'a low value upon trade unionism'.[24] He has also been saddled with the view that 'the trade-union movement ought to be the industrial arm of a political movement'.[25] Historically, these attitudes represent the received wisdom among the leaders of socialist parties; and, for a time, De Leon evidently shared them. But the emergence of the IWW in 1905 coincided with a dramatic shift in his ideas.[26] Thus he subsequently asserted that trade unionism was 'indispensable . . . for the emancipation of the Working Class';[27] and poured scorn on the argument (he attributed it to the Socialist party of America, with which his own party competed) that 'political action is all-sufficient to emancipate the Working Class'.

> The emancipation of the proletariat, that is, the Socialist Republic, can not be the result of legislative enactment. No bunch of office holders will emancipate the proletariat.[28]

The means of emancipation, he maintained, 'can only be the mass-action of the proletariat itself, "moving in," taking possession of the productive powers of the land'.[29] Political organization was not geared to this task. Indeed, 'in the act . . . of "taking and holding" the nation's plants of production, the political organization of the working class can give no help'—for the party's 'mission

20 De Leon, *Socialist Reconstruction of Society*, pp. 50–1.
21 Ibid. 60.
22 Ibid.
23 Ibid. 66.
24 Taft, 'Theories of the Labor Movement', p. 30.
25 Joll, *The Anarchists*, p. 201.
26 See McKee, 'Daniel De Leon: A Reappraisal', pp. 265–6.
27 De Leon, *Industrial Unionism*, p. 23.
28 De Leon, *As To Politics*, pp. 40–1.
29 Ibid. 41.

will have come to an end just before' emancipation was achieved in this way.[30]

There was a remarkable act of self-denial involved in De Leon's perception of the role of a true working-class party. Such a party, he argued, was concerned with the state and with the political power that the capitalist class exercised through the state. It sought to encroach on that power by electoral means. But it did so with the ultimate aim of abolishing the state as a source of power. Thus if its candidates gained the most complete election victory imaginable: 'Their work would be done by disbanding'[31]—in effect, by disbanding government. De Leon's point was that 'the goal of the political movement of labor is purely *destructive*'.[32] And that included self-destruction. 'The political movement of labor that, in the event of [electoral] triumph, would prolong its existence a second after triumph, would be . . . either a usurpation or the signal for a social catastrophe.'[33] The catastrophe would occur if the industrial movement was unprepared to step in and, fulfilling the constructive role within its capacity alone, take over the agencies of production and distribution.

Thus De Leon could say that 'the political movement is absolutely the reflex of economic organization'.[34] And he could write of the 'political arm of the Movement' as being 'worn away useless without the economic arm [the trade unions] is ready to second, to supplement, and, at the critical moment, to substitute it'.[35] Nevertheless, he still believed that political organization was not merely useful, but essential. Moreover, the effective role of the true working-class party—the Socialist Labor party—was not confined to the admittedly 'highly improbable' eventuality of a sweeping electoral victory.[36] Of far greater immediate importance was the SLP's political campaigning, in which it 'preaches the Revolution, teaches the Revolution'; and, in so doing, facilitated the ability of 'the IWW to recruit and organize its forces'.[37] The altruistic element was clear.

[30] De Leon, *Socialist Reconstruction of Society*, p. 60.
[31] Ibid. 58.
[32] Ibid. 57.
[33] Ibid. 58.
[34] Ibid. 54.
[35] De Leon, *Industrial Unionism*, p. 23.
[36] De Leon, *As To Politics*, p. 42.
[37] Ibid. 43.

The SLP man clings to political action because it is an *absolute necessity* for the formation of that organization—the IWW—which is both the embryo of the Workers' Republic and the physical force that the proletariat may, and in all likelihood will, need, to come to its own.[38]

Ultimately, in other words, political action was no less crucial than industrial: 'Without political organization, the labor movement cannot triumph.'[39]

For De Leon, moreover, there was another beneficial aspect to political action. As employed by the Chicago IWW, in particular, direct industrial action in America at that time involved a high level of violence. De Leon may not have opposed direct action on moral grounds, as has been argued,[40] but he certainly equated it with 'the methods of barbarism', and sharpened the point by contrasting political action as 'the civilized method'[41] because it involved the ballot—'a weapon of civilization'.[42] The importance of this weapon, in the United States at least, was that it might secure a 'peaceful solution of the great question at issue.'[43] In Europe a peaceful solution was beyond hope because of a persisting feudal emphasis on 'valor',[44] which meant that the ruling classes there would certainly put up a fight. There was, however, no such tradition in the United States which alone had reached the pinnacle ('full-orbed') of capitalist development.[45] The capitalist rulers of America, as a result of their upbringing, were basically bullies and cowards who were likely to cave in once the political battle had been won at the ballot box—provided, of course, that the party had raised 'the political temperature . . . to the point of danger' and that the trade unions stood ready to launch the general strike.[46] The American social revolution, in other words, could conceivably be bloodless.

De Leon's perception of the relationship between industrial and political action is unique among the leading syndicalist spokesmen. On the other hand, as a peculiarly American perception, it seems to have found at least a partial echo in the thinking of the Chicago

[38] De Leon, *As To Politics*, p. 44. Emphasis added.
[39] De Leon, *Socialist Reconstruction of Society*, p. 70.
[40] See Brissenden, *The I.W.W.*, p. 236.
[41] De Leon, *As To Politics*, p. 41.
[42] De Leon, *Socialist Reconstruction of Society*, p. 62.
[43] Ibid. 61.
[44] Ibid. 68.
[45] Ibid. 55.
[46] Ibid. 68–9.

IWW, conventionally identified as rigidly Anarcho-syndicalist on the issue. For in a speech delivered in 1911, and later published as a pamphlet, William Haywood conceded a role to 'political action' and 'the power of the ballot' which leans plainly towards De Leon's position rather than that of the French syndicalists.[47] In any case, it was in this respect that De Leon contributed an unequivocally distinctive element to One Big Unionism. Its other distinctive element was the trade union structure embodied in the IWW, especially in its Chicago version. This evolved (at least on paper) into an extraordinarily complex and highly centralized organization;[48] and, in their practice, the leaders of the Chicago IWW sought to exercise precisely the kind of central control it implied.[49] In both aspects, it was totally at odds with the expressed organizational ideals of the Anarcho-syndicalists.

GUILD SOCIALISM

French Anarcho-syndicalism and American One Big Unionism were both essentially movements of trade unionists, with intellectuals playing a relatively subordinate role. The reverse was true in the case of English Guild Socialism. The difference is perhaps most evident in the meticulous care with which the contours of the future society were mapped in the writings of Guild Socialism—'the Platonic idea' of syndicalism, as Schumpeter happily described it.[50]

Guild Socialism, which emerged shortly before the first world war, found its most influential advocates in S. G. Hobson and G. D. H. Cole. Unlike other syndicalists, they did not focus exclusively on the manual worker. They envisaged that workers' control of industry would be exercised through associations which (unlike conventional trade unions, as they understood them) would represent all the employees in an industry, both manual and non-manual. Their recognition of non-manual employees reflects the Guild Socialists' sensitivity to complexities ignored by the Anarcho-syndicalists and the One Big Unionists.[51] Another reflection is the way in which they handled the issue of the state.

[47] Quoted in Kornbluh, *Rebel Voices*, p. 49.
[48] See e.g. Brissenden, *The I.W.W.*, App. III.
[49] See Renshaw, *The Wobblies*, ch. 6.
[50] Schumpeter, *Capitalism, Socialism and Democracy*, pp. 339–40.
[51] See Cole, *Chaos and Order in Industry*, p. 45.

The Guild Socialists did not accept the other syndicalists' vision of a new society in which producers' organizations, the trade unions or guilds, stood alone in the field. They were sensitive, like collectivist socialists, to the interests of consumers, and conscious that a guild might overstep the mark of fair and reasonable behaviour in this respect. This, of course, is where the state, or something like it, comes in. Hobson, in any case, believed that there were non-economic functions which only the state could perform. Apart from this, in his earliest writings he referred to the new society as involving 'co-management' between the guilds and the state, with the state as 'the final arbiter', and 'probably even supreme'.[52] The uncertainty of his 'probably' seems to have been resolved by 1920 when he spoke of guilds obtaining a 'charter' from the state, and being subject to state 'intervention' in the event of a breach.[53] There is, however, one significant qualification to be noted in relation to this unsyndicalist recognition of the state. The state Hobson had in mind was itself subject, in its administrative aspect, to workers' control by way of a 'Civil Service Guild'.[54]

Cole was less ambivalent in the matter of the state. Initially, it is true, he welcomed Hobson's notion of 'co-management' and thought of the trade union as 'the future partner of the State in the control of industry', with the latter retaining 'supreme control' in order to obviate 'Guild profiteering'.[55] But by 1920, Cole was flatly against the state. The guild system, he still thought, required a measure of central 'co-ordination', but that was best provided in the form of 'self-co-ordination'.[56] The outcome was the 'National Commune', heading a highly decentralized system of regional and local communes, each consisting mainly of members drawn from guilds and consumer associations. The National Commune itself would have 'few direct administrative functions', and would in general be 'a much less imposing body' than the state with its 'huge machinery of coercion and bureaucratic government'.[57] Nevertheless, Cole was still proposing to constrain the trade unions in a way never contemplated by the Anarcho-syndicalists and the One Big Unionists.

[52] Hobson, *National Guilds*, pp. 132–3, 150, 259, 263.
[53] Hobson, *National Guilds and the State*, pp. 126, 292.
[54] Ibid. 320.
[55] Cole, *The World of Labour*, pp. 51, 363–9, 406.
[56] Cole, *Guild Socialism Re-stated*, p. 124.
[57] Ibid. 136–7.

The same note of moderation flavoured the Guild Socialists' approach to the transition from private ownership to workers' control. They shared the belief of the other syndicalists in the critical importance of trade unions and the strike: 'National Guilds can never be realised save by economic action and by industrial associations'.[58] But then came the qualifications. There was no need for the general strike; and, hopefully, no need for violence. Hobson might employ the language of warfare with his references to 'the labour army',[59] but the army's strategy for destroying capitalism involved a process that was to be gradual, piecemeal, and essentially peaceful. Industrial action was to be used, selectively and on a co-ordinated basis, to gain increasing degrees of control in particular enterprises, and eventually 'complete control'.[60] Cole entitled this process 'encroaching control', and described it as 'a policy directed to wresting bit by bit from the hands of the possessing classes the economic power which they now exercise, by a steady transference of functions and rights from their nominees to representatives of the working-class'.[61]

Cole also added another significant dimension to the process. In some key industries, he argued, the path to workers' control would involve government action in the form of nationalization. 'State management' did not in itself change the essential nature of the worker's position; but, 'by clearing the private owner out of the way', it would make the achievement of workers' control easier in particular cases.[62] In this respect, political action had a role to play. But, much like De Leon, Cole insisted that it was no more than 'an important secondary function', which served 'to ease and smooth a transition' that depended primarily on direct action: 'It is the economic, rather than the political, power of the workers that will avail to overthrow capitalism'.[63]

The Guild Socialists, as syndicalists, were distinguished by a sense of caution and an air of patience. Yet they were still revolutionaries, as Clegg has pointed out, in that they 'sought a

[58] Hobson, *National Guilds and the State*, p. vi.

[59] Hobson, *National Guilds*, p. 106.

[60] Ibid. 107–8.

[61] Cole, *Guild Socialism Re-stated*, p. 196; and see Cole, *Chaos and Order in Industry*, pp. 117, 154–5.

[62] Cole, *Guild Socialism Re-stated*, pp. 205–6; and see Cole, *Self-government in Industry*, pp. 216–20.

[63] Cole, *Guild Socialism Re-stated*, p. 180.

complete transformation of society'.[64] And they have usually been depicted as expecting a transformation that was, above all, non-violent.[65] That position was certainly Hobson's; and, for a time, Cole's also. By 1920, however, Cole had come to a different conclusion. He now thought that 'a revolutionary element', entailing violence, 'is unavoidable in any "thorough" policy of social transformation'.[66] The best that could be hoped for was that enough might be achieved beforehand—by 'constitutional means, industrial or political'—to ensure that the 'unconstitutional "revolution"', when it came, would be 'as little as possible a civil war and as much as possible a registration of accomplished facts'.[67] In other words, even if violence were inevitable, it was still possible and desirable for the demolition of capitalism to be achieved largely by whittling it away, rather than blowing it up. And trade unionism remained the key to this process. It was, as Cole remarked, 'the successor of Capitalism as well as its destroyer'.[68]

WHOSE INTERESTS?

Syndicalism had no truck with the narrow conception of trade union responsibility favoured by pluralists. Each of its three variants was geared to the total overthrow of capitalism—and thus to the general realization of workers' control. This implies a conception of responsibility that extends far beyond the sectional interests of a particular trade union's membership. There was, however, disagreement about the precise boundaries involved.

For Anarcho-syndicalists, the boundaries were set by the interests of the working class as a whole. This was reflected in their insistence on the reality of a class war between the proletariat and the bourgeoisie, an idea that one of them claimed as 'the beginning and the end of syndicalism'.[69] The same idea, with its emphasis on the working class at large, was central, too, for the One Big Unionists. The preamble to the Chicago IWW's constitution, in which that was formally stated, also spoke of the IWW as

64 Clegg, *A New Approach To Industrial Democracy*, p. 12.
65 See Glass, *The Responsible Society*, pp. 5, 24.
66 Cole, *Guild Socialism Re-stated*, p. 182.
67 Ibid. 187.
68 Cole, *The World of Labour*, p. 369.
69 Lagardelle, cited in Ridley, *Revolutionary Syndicalism in France*, p. 100.

upholding 'the interest of the working class'.[70] The Guild Socialists, on the other hand, had a somewhat wider conception of responsibility in mind.

In all three versions of syndicalism workers' control, or self-government in industry, was the heart of the matter. However, the Guild Socialists, unlike the others, gave a carefully specific answer to the question of *who* were to become self-governing. Their answer was: *all* employees, whatever their function and whatever their status. It followed that the primary union responsibility was to both manual and non-manual employees, including those at professional and managerial levels. It was precisely because they wanted to make this clear that the Guild Socialists adopted the term 'guild': 'We could not use the word "union" because that implies a combination of manual workers—proletarians; whereas ... we have predicated ... a combination of all the industrial and commercial functions—wage, salariat, administration.'[71] By the same token, the Anarcho-syndicalists and the One Big Unionists used the terms 'union', 'proletariat', 'working class', precisely because they were thinking exclusively in terms of manual workers, and conceived union responsibility in the same terms. Anarcho-syndicalism, indeed, 'preached class war in its purest form' with a future to be won by the manual workers, entirely without help from non-proletarians—and so to Pouget's final stage in which 'all parasites will be eliminated and only the working class will survive'.[72] Malatesta, a more traditional anarchist, was moved to comment that the 'anarchist revolution', properly speaking, 'far exceeds the interests of a single class'.[73] But of all the syndicalists, only the Guild Socialists acknowledged his point.

INTERESTS IN WHAT?

Trade unions had two supreme purposes, one sequential to the other, for all the syndicalists. The first was to overthrow the capitalist system. The second was to control the productive resources at least (to acknowledge limits foreseen only by the Guild

[70] Quoted in Brissenden, *The I.W.W.*, p. 352.
[71] Hobson, *National Guilds*, p. 275.
[72] Ridley, *Revolutionary Syndicalism in France*, p. 268.
[73] In Woodcock, *The Anarchist Reader*, p. 225.

Socialists) of the new society. De Leon provided a vivid summation: 'the Industrial Union is at once the battering ram with which to pound down the fortress of Capitalism, and the successor of the capitalist social structure itself'.[74] In addition, however, there was another layer of purpose—in effect a third purpose. And it was the key to the achievement of the two supreme purposes. Workers had to be convinced of both the legitimacy and the feasibility of the 'battering ram' and 'successor' roles.

Each of the two supreme purposes entailed different functions on the part of trade unions. The 'battering ram' involved disciplined use of the strike, whether in the form of a climactic general strike or as part of a strategy of encroaching control; and syndicalists generally thought of this as an essentially *military* function requiring stern resolve and high morale. The 'successor' purpose, on a syndicalist interpretation, involved an *administrative* function which (as the Guild Socialists at least were aware) required access to technical expertise.

The third purpose, convincing workers of their historical mission, entailed a function that was subsidiary, but utterly essential, to both the military and the administrative functions. This was the *educational* function which the syndicalists attributed to trade union organization.

Through the trade unions workers were to be educated, above all, in the sense of being persuaded that they held their fate in their own hands. Thus Fernand Pelloutier wrote of the need 'to prove to the working mass . . . that a government by itself and for itself is possible', and of 'instructing it in the need for revolution'.[75] In addition, although this was in general given much less emphasis, the unions were also to educate the workers in the sense of equipping them to govern themselves in their workplaces.

Pelloutier was one syndicalist who conceived the trade unions' educational function in formal didactic terms, involving organized programmes of instruction. He encouraged the local trade union organizations he was associated with to become 'centres of study where the proletariat could reflect on their condition, [and] unravel the elements of the economic problem so as to make themselves capable of the liberation to which they have the right'.[76] Most

[74] De Leon, *Industrial Unionism*, p. 65.
[75] Quoted in Horowitz, *Radicalism and the Revolt Against Reason*, p. 250.
[76] Quoted in Joll, *The Anarchists*, p. 181.

syndicalists, however, thought of the educational function as being less a matter of formal instruction than of lessons taught to workers through their involvement in union affairs and activities. 'Experience', Lagardelle remarked, was 'the school of syndicalism'.[77] Participation in strikes was usually seen as a particularly important form of such experience, especially by Anarcho-syndicalists and One Big Unionists. Monatte made the point: 'it is by the strike that the masses receive their revolutionary education, that they understand their strength and that of the enemy, that they take confidence in their power and learn to be audacious.'[78] The Guild Socialists, characteristically, were more inclined to look to the future and stress administrative competence rather than revolutionary audacity.

Responsibility is the best teacher of self-reliance: self-government in the Trade Union has done wonders for the workers . . . But with the gradual extension of Trade Union competence to cover more and more of the industrial field, the lessons it will be able to afford will be of infinitely greater value. In controlling industry, democracy will learn the hard lesson of self-control and the harder lesson of controlling its rulers.[79]

Variations of emphasis aside, however, the syndicalists were at one in appreciating the centrality of the trade union's educational function. De Leon spoke for both the American and the French brands when he asserted that 'the facilities enjoyed by the Trades Union as an Academy for drilling its membership in the two essentials for the emancipation of their class—discipline and class-consciousness—are matchless'.[80] Speaking for the Guild Socialists, Cole was more laconic, but his point was essentially the same: 'the Trade Unions are the most powerful instruments for the education of the people'.[81]

[77] Lagardelle, cited in Ridley, *Revolutionary Syndicalism in France*, p. 263.
[78] In Woodcock, *The Anarchist Reader*, p. 218.
[79] Cole, *The World of Labour*, pp. 28–9.
[80] De Leon, *Industrial Unionism*, p. 18.
[81] Cole, *The World of Labour*, p. 411.

4

THE MARXIST-LENINISTS

Trade Unions as Party Instruments

The Leninist contribution to the broad stream of Marxist thought incorporates two different answers to the question of trade union purpose. For Lenin, unlike Marx and Engels, confronted the question not only in relation to capitalist society, but also in the context of a new socialist society. His views on trade unions under socialism are considered in Chapter 6. What I describe as the Marxist-Leninist vision of purpose is thus concerned solely with trade unions under capitalism.

There is not a great deal about trade unions in the torrent of words that poured from the pens of Marx and Engels. Indeed, in the case of works of primary importance written by Marx, trade unions figure explicitly—and then in a minor way—in only four: *The Poverty of Philosophy* (1847), *The Communist Manifesto* (1848), *Wages, Prices and Profit* (1865), and the 'sixth chapter' omitted from *Capital* (vol. 1) and unpublished until 1933. Otherwise, Marx's surviving comments on unions are confined to letters, speeches, reports, newspaper articles, a circular, and some conference resolutions he drafted. But there can be no doubt about his interest in trade unionism, especially during the 1860s and early 1870s. In company with some union leaders he helped launch the International Working Men's Association (later known as the First International) in 1864, and took a leading part in its affairs for the next eight years: 'I am in fact the head of the thing', he told Engels in 1865.[1] He was also instrumental in securing its effective demise (by having its executive body transferred from London to New York) as a means of averting a takeover by Michael Bakunin. His recorded comments on trade unions help explain the devotion of so much energy to the First International, though they do not provide as clear a picture of his ideas on trade union purpose as one might wish.

[1] Quoted in Lapides, *Marx and Engels on the Trade Unions*, p. xvi.

Lenin, although deeply interested in trade union organization, was less intimately involved with it. On the other hand, trade union issues bulk rather larger in his writings, and his position on trade union purpose is stated with more precision.

WHOSE INTERESTS?

Implicit in everything that Marx and Engels wrote about trade unions was the belief that the true purpose of unions was to serve the interests of the working class as a whole, and not merely the interests of those workers who happened to comprise their membership. Marx made the point briefly in 1865 when he ascribed to trade unions the task of 'using their organized forces as a lever for the final emancipation of the working class'.[2] The following year, in the course of a long resolution he drafted for the congress of the First International, he left no doubt about the extent of their responsibility.

> By considering themselves *champions and representatives of the whole working class*, and acting accordingly, the trade unions must succeed in rallying round themselves all workers still outside their ranks. They must carefully safeguard the interests of the workers in the poorest-paid trades . . . They must convince the whole world that their efforts are far from narrow and egoistic, but on the contrary, are directed towards the emancipation of the down-trodden masses.[3]

Lenin echoed the point when he remarked that workers learned, through trade union action, 'to make war on their enemies for the liberation of the whole people, of all who labour'.[4]

INTERESTS IN WHAT?

Marx and Engels

It was Engels who established, in print, the starting-point of Marx's position on the second dimension of trade union purpose under capitalism. In his study *The Condition of the Working Class in England*, first published in 1845, Engels fixed on the elimination of competition among workers for jobs as the central goal of trade unions, and remarked on its significance.

[2] In Clarke and Clements, *Trade Unions under Capitalism*, p. 55.
[3] Quoted in Lozovsky, *Marx and the Trade Unions*, p. 18. Emphasis added.
[4] Lenin, *Collected Works*, iv. 317.

But what gives these Unions . . . their real importance is this, that they are the first attempt of the workers to abolish competition. They imply the recognition of the fact that the supremacy of the bourgeoisie is based wholly upon the competition of the workers among themselves. And precisely because the Unions direct themselves against the vital nerve of the present social order . . . are they so dangerous to this social order. The working-men cannot attack the bourgeoisie . . . at any sorer point than this.[5]

Not long afterwards, Marx pointed to another layer of purpose when he linked the issue of competition to the issue of wages.

But the maintenance of wages . . . unites [workers] in a common thought of resistance—combination. Thus the combination [trade union] always has a double aim, that of stopping competition among the workers, so that they can carry on a general competition with the capitalists.[6]

The point was reaffirmed in their joint work, *The Communist Manifesto*, which referred to the way workers 'club together to keep up the rate of wages'.[7]

On the other hand, the wages issue (to which Marx later added working hours) was merely the 'immediate aim of the trade unions'.[8] This immediate aim was transcended by a much larger purpose. The trade unions had a 'great historical mission'; they had become 'the focal points for the organization of the working class'.[9] Their destiny was to provide, as Marx later put it, the 'lever of proletarian revolution'.[10] Their supreme purpose, in other words, was nothing less than the destruction of capitalism.

With the greater purpose there came a more momentous function. 'In addition to their orginal tasks, the trade unions must now learn how to act consciously as focal points for organizing the working class in the greater interests of its complete emancipation'.[11] 'Organizing' in this context meant, above all, that the trade unions were to play a leading part (perhaps *the* leading part) in the historical process whereby the workers would become class-conscious—that is, aware that they shared, as a class, an interest

[5] Engels, *The Condition of the Working Class in England*, p. 226.
[6] In McLellan, *Karl Marx: Selected Writings*, p. 213–4.
[7] Ibid. 228.
[8] Quoted in Lozovsky, *Marx and the Trade Unions*, p. 16.
[9] Ibid. 17.
[10] Quoted in Rubel, *Marx*, p. 88.
[11] Quoted in Lozovsky, *Marx and the Trade Unions*, p. 18.

in destroying capitalism which far outweighed their sectional interests. The leading function of trade unions was thus to raise the consciousness of the workers. It was, in a word, an educational function.

Again, it was Engels who first explicitly advanced this idea in print. He did so by applying the term 'school' to both unions and strikes (which he regarded as inseparable).

> These strikes . . . are the military school of the working-men in which they prepare themselves for the great struggle which cannot be avoided . . . And as schools of war, the Unions are unexcelled. In them is developed the peculiar courage of the English.[12]

Twenty-four years later, Marx employed the same term ('schools') to make the same kind of point.

> Trade unions are the schools of socialism. It is in trade unions that workers educate themselves and become socialists, because under their very eyes and every day the struggle with capital is taking place.[13]

Trade unions discharge their educational function, as this quotation implies, primarily through the experiences they provide for their members. Their strikes, for example, in particular cases 'revolutionized the industrial proletariat'.[14] In other words, unions help shape an environment in which the worker 'becomes socialist without noticing it'.[15]

There can be no doubt that Marx placed great store in the role of the trade unions in this respect. He stressed more than once that they were 'of the utmost importance'.[16] Again: 'If trade unions have become indispensable for the guerilla fight [about wages etc.] between Capital and Labour, they are even more important as organized bodies to promote the abolition of the very system of wage labour'.[17] But the achievement of the trade unions' historic purpose was hampered by one fact and qualified by another. First, there was the problem of self-recognition. The trade unions themselves, Marx noted in 1866, 'have not yet completely realized their power to attack the very system of wage slavery'.[18] They had

[12] Engels, *The Condition of the Working Class in England*, pp. 230–1.
[13] In McLellan, *Karl Marx: Selected Writings*, p. 538.
[14] In Lapides, *Marx and Engels on the Trade Unions*, p. 53.
[15] In McLellan, *Karl Marx: Selected Writings*, p. 538.
[16] Ibid. 585; see also ibid. 521.
[17] Quoted in Lozovsky, *Marx and the Trade Unions*, p. 16.
[18] Ibid. 17.

acquired such power 'without being aware of it'.[19] As a result, they still confined themselves 'to a guerilla war against the effects of the existing system, instead of simultaneously trying to change it'.[20] Their awareness had to be raised before their true potential could be realized.

Secondly, the importance of the trade unions was qualified in that, unlike the syndicalists of the future, neither Marx nor Engels believed that trade unionism, on its own, could accomplish the overthrow of capitalism. As Engels early remarked, 'something more is needed than Trades Unions and strikes to break the power of the ruling class'.[21] It is true that in a speech to German unionists, Marx once inclined toward a syndicalist stance when he insisted that 'trade unions ought never to be attached to a political association'.[22] But that observation (as Lenin was to argue)[23] seems to have been inspired by the specific exigencies of German politics in 1869 rather than by a general theoretical position. Certainly, Marx had earlier lamented the way unions 'kept aloof from social and political movements';[24] and, subsequently, had advocated working-class political action in the belief that 'to get workers into the parliaments is equivalent to a victory over the governments'.[25] Later still, Engels extolled the achievements of conventional political action in terms which effectively dismissed the unions' role.[26] Moreover, almost fifty years before, he and Marx in *The Communist Manifesto* had implicitly foreshadowed another line of reasoning pointing in the same direction, when they spoke of Communists as having 'over the great mass of the proletariat the advantage of clearly understanding the line of march, the conditions, and the ultimate general results of the proletarian movement'.[27] It was Lenin, however, who developed these hints into a coherent theory of trade unionism's limitations in the struggle against capitalism.

[19] Quoted in Lozovsky, *Marx and the Trade Unions*, p. 16.
[20] In Clarke and Clements, *Trade Unions under Capitalism*, p. 55.
[21] Engels, *The Condition of the Working Class in England*, p. 226.
[22] In McLellan, *Karl Marx: Selected Writings*, p. 538.
[23] See Lenin, *Collected Works*, xiii. 466–7.
[24] Quoted in Lozovsky, *Marx and the Trade Unions*, p. 17.
[25] Quoted in Rubel, *Marx*, p. 95.
[26] See Tucker, *The Marx–Engels Reader*, pp. 556–73.
[27] In McLellan, *Karl Marx: Selected Writings*, p. 231.

Lenin and Others

The most elaborate statement of Lenin's views on trade unions under capitalism occurs in *What is to be Done?*, published in 1902. This long pamphlet revolves around an attack on 'Economism', a doctrinal trend within the Russian Social-Democratic Workers' Party which veered towards syndicalism by emphasizing trade union organization and industrial action at the expense of their political counterparts.[28] The heart of this position, and the focus of Lenin's attack, was the so-called 'spontaneity' thesis—the proposition that the achievement of revolutionary class-consciousness by the workers depended, above all, on their own direct experience and on their own organizations, the trade unions.

In *What is to be Done?* Lenin depicted trade union purpose in terms which gave it a political dimension but, at the same time, confined it essentially to the area of what Marx, as we have seen, had called 'immediate aims'. Thus, for Lenin, there was on the one hand the 'economic struggle' against employers, 'necessarily a trade-union struggle', in which workers tried to secure 'better terms in the sale of their labour-power'.[29] On the other hand, there was 'trade-union politics' which concerned workers' efforts to secure government measures alleviating their conditions—measures, however, which stopped short of the cause of those conditions in that they did nothing to 'abolish . . . the subjection of labour to capital'.[30]

Lenin's point was that the purposes of trade unionism were limited to reform. They did not touch the fundamental issue. And whereas Marx had once assured the unions of a 'great historical mission', as 'focal points' of working-class organization, Lenin would have none of it. For him, the unions' problem of self-recognition (which Marx seemed to imply was temporary) was a permanent condition. Lenin had no doubt that the 'mission' Marx had envisioned belonged not to the trade unions, but to the party. It was the true focal point.

The party, he argued, stood for 'Social-Democratic politics' which, unlike 'trade-unionist politics', was concerned above all with fundamentals.[31] 'Social-Democracy leads the struggle of the

[28] See Hammond, *Lenin on Trade Unions and Revolution*, ch. 3.

[29] Lenin, *Collected Works*, v. 404.

[30] Ibid. 387.

[31] Ibid. 408.

working class, not only for better terms for the sale of labour-power, but for the abolition of the social system that compels the propertyless to sell themselves to the rich'.[32] Furthermore, the party ('Social Democracy') had its priorities straight: 'it subordinates the struggle for reforms, as the part to the whole, to the revolutionary struggle for freedom and for socialism'.[33] In other words, unlike trade unionism, the party was an innately revolutionary force.

The 'Economists', because of their 'nursemaid concern' for the workers[34] and their obsession with 'merely trade-union work',[35] were in fact 'striving *to degrade* Social-Democratic politics to the level of trade-union politics'.[36] The narrow scope of trade union politics was precisely a function of the limited horizons of the proletariat as a class.

> The history of all countries shows that the working class, exclusively by its own effort, is able to develop only trade-union consciousness, i.e., the conviction that it is necessary to combine in unions, fight the employers, and strive to compel the government to pass necessary labour legislation, etc.[37]

Thus, it was futile to expect the spontaneous generation of class-consciousness among workers through their trade unions: 'there can be no talk of an independent ideology formulated by the working masses themselves in the process of their movement'.[38] In this argument, Lenin was not only denying the 'Economists', with their 'subservience to spontaneity',[39] but also (although he made no mention of this) what Marx and Engels had seemed to imply when they described trade unions as 'schools of war' and 'schools of socialism'.

The other side of Lenin's argument followed inexorably. 'Class political consciousness can be brought to the workers *only from without*, that is, only from outside . . . the sphere of relations between workers and employers'.[40] The reason for this was that such consciousness depended upon 'political knowledge'[41] (in Karl Kautsky's phrase, 'profound scientific knowledge').[42] And that knowledge was generated by 'educated representatives of the propertied classes, by intellectuals',[43] not by the workers as such.

32 Lenin, *Collected Works*, v. 400.
33 Ibid. 406.
34 Ibid. 433.
35 Ibid. 400.
36 Ibid. 405.
37 Ibid. 375.
38 Ibid. 384.
39 Ibid. 378.
40 Ibid. 422.
41 Ibid.
42 Quoted, ibid. 383.
43 Ibid. 375.

Certainly, individual workers of talent might have a hand in this; but in so far as some did, it was 'not as workers but as socialist theoreticians, as Proudhons and Weitlings'[44]—in short, as intellectuals.

In the absence of the knowledge provided by socialist theoreticians, left to the workings of 'spontaneity', the workers were easy prey:

> the *spontaneous* development of the working-class movement leads to its subordination to bourgeois ideology . . . for the spontaneous working-class movement is trade-unionism . . . and trade-unionism means the ideological enslavement of the workers by the bourgeoisie.[45]

Those who proposed to leave the working class and the trade unions to their own devices were, quite simply, delivering them into the hands of the capitalist ruling class. 'Trade-unionist politics of the working class', Lenin insisted, 'is precisely *bourgeois politics* of the working class.'[46]

The lesson was plain. It was necessary 'to *combat spontaneity*, *to divert* the working-class movement from this spontaneous, trade-unionist striving to come under the wing of the bourgeoisie, and to bring it under the wing of revolutionary Social-Democracy'.[47] This task required leadership. Above all, it required 'wise men'; and 'by "wise men" . . . I mean *professional revolutionaries*'.[48] Such men needed a breadth of vision which was well beyond the capacity of one preoccupied with the minutiae of trade union politics:

> the Social-Democrat's ideal should not be the trade-union secretary, but *the tribune of the people* . . . who is able to take advantage of every event, however small, in order to set forth *before all* his socialist convictions and his democratic demands, in order to clarify for *all* . . . the world-historic significance of the struggle for the emancipation of the proletariat.[49]

And Lenin reaffirmed the contrast in roles with a cutting description of an incompetent Social-Democrat ('not a revolutionary, but a wretched amateur!') who, among other things, 'resembles a trade-union secretary more than a spokesman of the people'.[50]

It was the party, of course, that contained the wise men, the professional revolutionaries, who alone were capable of acting as

[44] Ibid. 384 n. [45] Ibid. 384. [46] Ibid. 426. [47] Ibid. 384–5.
[48] Ibid. 464. [49] Ibid. 423. [50] Ibid. 466.

the 'vanguard' of the proletariat.[51] There was, certainly, a place for workers among them. Indeed,

> our very first and most pressing duty is to help train working-class revolutionaries who will be on the same level *in regard to Party activity* as the revolutionaries from amongst the intellectuals (we emphasise the words 'in regard to Party activity', for, although necessary, it is neither so easy nor so pressingly necessary to bring the workers up to the level of the intellectuals in other respects).[52]

But, for the 'worker-revolutionary' to fit the vanguard role, he had to cease being a worker and, instead, 'become a professional revolutionary'.[53] The point, in the end, was that workers, as such, had *nothing* to offer the vanguard: that was the ultimate implication of Lenin's attack on the spontaneity thesis.

Richard Hyman has argued that Lenin's spurning of the spontaneity thesis in *What is to be Done?* 'accords ill with certain of his earlier and later writings', in which he echoed Marx and Engels by placing 'considerable emphasis' on the 'potential of trade union struggle in raising workers' consciousness'.[54] This argument, in any case somewhat cloudy in its terms, neglects one relatively clear-cut consideration—the dominating role which the Lenin of *What is to be Done?* attributes, in plain contrast to Marx and Engels, to a party of professional revolutionaries. Hyman cites, or quotes from, six of Lenin's other works in support of his argument. But in four of them (the two exceptions involved formal celebrations, one in 1910 and the other in 1917, of the 1905 revolution) the expressed support for the spontaneity thesis, which Hyman highlights, is accompanied by one or more references that stress the role of the party and, effectively, reject the spontaneity thesis as the 'Economists' understood it. One example makes the point. In 'The Reorganisation of the Party' (1905) the statement, 'The working class is instinctively, spontaneously Social-Democratic' (which Hyman quotes in support of his argument),[55] is followed immediately by: 'and more than ten years of work put in by Social-Democracy [the party] has done a great deal to *transform this spontaneity into consciousness*'.[56] The skill of the politician who deals, of necessity, in ambiguity, but nevertheless has a firm

[51] Lenin, *Collected Works*, v. 426. [52] Ibid. 470. [53] Ibid. 472.
[54] Hyman, *Marxism and the Sociology of Trade Unionism*, p. 41.
[55] Quoted, ibid. p. 42.
[56] Lenin, *Collected Works*, x. 32. Emphasis added.

position, is evident in this. The role of the party, as spelt out more bluntly in *What is to be Done?*, is reaffirmed—and with it the rebuttal of the spontaneity thesis.

At the same time, there is no question that Lenin attached great significance to trade unions. The point, however, is that their significance as *independent* entities was limited to the 'economic struggle'. When it came to the more momentous 'revolutionary struggle', they had significance only as instruments of the party. It was thus up to the party, by 'generalizing the trade-union struggle', to strengthen 'the link between the Russian trade-union movement and socialism'.[57] For it was only under the party's guidance that the operative purpose of the trade unions would be extended beyond mere reformism to include the revolutionary objective of eliminating capitalist exploitation. There was no doubt about the importance of the unions in this connection; but, equally, there was no doubt either about their subsidiary status. 'Trade-union organizations, not only can be of tremendous value in developing and consolidating the economic struggle, but can also become a very important *auxiliary* to political agitation and revolutionary organization.'[58]

Other influential strains of Marxist thought have echoed Lenin's perception of the trade union as subsidiary to 'political agitation and revolutionary organization'. Thus Rosa Luxemburg agreed that trade unions were confined by their nature to 'the economic guerilla war';[59] and this, while important for the immediate interests of the proletariat, condemned the unions (in her famous phrase) to 'a sort of labour of Sisyphus'.[60] She poured scorn on 'the illusion of equality' between party and unions.[61] The true 'relation of the trade unions to Social Democracy is that of the part to the whole'.[62] The superiority of the party could scarcely have been put more brutally. Nevertheless, Luxemburg vigorously denied Lenin's conception of a highly centralized, directorial party. Certainly, the party was the guiding spirit in relation to the trade unions, but she conceived its role as being rather less imperious.[63] For Luxemburg, unlike Lenin, thought there was something in the spontaneity thesis. She apparently believed that, in the end, the proletariat had

[57] Ibid. v. 492. [58] Ibid. 456–7. Emphasis added.
[59] Luxemburg, *Selected Political Writings*, p. 262.
[60] Ibid. 105. [61] Ibid. 266.
[62] Ibid. 253. [63] See ibid. 290.

more to learn from its own experience than from the book-learning of middle-class intellectuals. The 'role of director', she argued, properly belonged to 'the collective ego of the working class, which insists on its right to make its own mistakes and to learn the historical dialectic by itself'.[64] Moreover, the 'errors made by a truly revolutionary labour movement are historically infinitely more fruitful . . . than the infallibility of the best of all possible "central committees" '.[65]

Antonio Gramsci was even more emphatic about the subordinate status of trade unions. Lenin and Luxemburg at least thought of them as the most significant form of purely working-class organization. They were not even that for Gramsci. Immeasurably more important, he thought, was the 'Factory Council', elected directly by workers on the job: this was 'the model of the proletarian State' and provided 'the basis of a new representative system—the system of Councils'.[66] The factory council, it followed, was 'an institution that cannot possibly be confused with, coordinated with or subordinated to the trade union; on the contrary . . . it effects radical changes in the form and structure of the unions'.[67] It would do so by instilling in workers 'a producer's mentality—the mentality of a creator of history', which the workers would 'carry . . . into the trade unions', and thereby 'renew fundamentally' existing trade union organs.[68] The council, in short, would act as 'a reagent dissolving the union's bureaucracy and bureaucratism'.[69] On the other hand, the factory councils' relationship with the *party* was a different matter. For Gramsci, unlike Luxemburg, viewed the party in Leninist terms. It was 'the motor centre' which should 'coordinate and centralize in its central executive committee the whole of the proletariat's revolutionary action'.[70] Accordingly, it was the party's factory cells which would 'organize the setting-up of Factory Councils', and 'develop the propaganda needed to conquer the unions'.[71]

Anton Pannekoek, doyen of the so-called 'Council Communists',[72] pushed the subordination of the trade unions to the ultimate extreme—extinction. He also formulated a radically different

[64] Luxemburg, *Selected Political Writings*, p. 306. [65] Ibid.
[66] Gramsci, *Selections from Political Writings 1910–1920*, pp. 100, 263.
[67] Ibid. 257.
[68] Ibid. 101, 298. [69] Ibid. 266. [70] Ibid. 195.
[71] Ibid. [72] See McLellan, *Marxism After Marx*, p. 171.

conception of party. The key to his position lies in his assumption that the total destruction of capitalism was unattainable—at least in developed societies—by 'an ignorant mass' under the leadership of a Leninist party: this end could be achieved only 'if the workers themselves, the entire class, understand the conditions, ways and means of their fight; when every man knows from his own judgement what to do'.[73] Both the Leninist party and the trade unions were 'old forms of organization', appropriate to the earlier period of 'expanding capitalism', but useless relics in a time of 'declining capitalism'.[74] New forms of organization, 'a new workers' movement', were required.[75] In the first place, if 'the final defeat of capitalism' was to be achieved, 'the masses' had to 'organize their power in the factories and workshops' by way of 'workers' councils'.[76] The councils, by producing 'a community that is increasingly close-knit', would 'prove that they [the masses] are not as incapable of creating the revolution as was supposed'.[77] In the second place, alongside the councils, 'the natural organs of education and enlightenment' were neither the trade unions nor the party, but rather 'work groups, study and discussion circles . . . formed of their own accord and . . . seeking their own way'.[78] These work groups, Pannekoek conceded, might be described as parties—so long as their extreme variation from the Leninist version was acknowledged.[79] They had 'one "primordial" task: to go out and speak to the workers'; and, in doing so, 'to spread insight and knowledge . . . and . . . enlighten the minds of the masses'.[80] By 'building up . . . [the] spiritual power' of the working class in this way, the work groups-cum-parties would play an essential role in the 'self-liberation' of the class.[81] For their part, the workers' councils were also indispensable, but for a different reason. They were to be the actual destroyers of capitalism. 'The workers' councils are the organs for practical action and fight of the working class.'[82] There was, of course, no place for the trade union in any of this.

It so happens that a younger, pre-workers' council, Pannekoek had regarded trade unions more kindly. Then, as 'the only genuinely working-class organizations', he charged them with 'the

[73] Pannekoek, *Lenin as Philosopher*, p. 76.
[74] Quoted in McLellan, *Marxism After Marx*, p. 173.
[75] Quoted in Bricianer, *Pannekoek and the Workers' Councils*, p. 261.
[76] Quoted, ibid. 265. [77] Ibid. [78] Quoted, ibid. 262.
[79] See ibid. 263. [80] Quoted, ibid. 267. [81] Ibid. [82] Ibid.

enormous task of [providing the] moral education required to transform the weak worker into the conqueror of capitalism'.[83] In doing so, he effectively specified a function which Lenin himself had envisaged for the trade unions in their role of 'a very important auxiliary' to the party.[84] Lenin discerned two functions altogether. One, in effect, was a military function which would come into play as an element of the revolution itself. Thus he referred to 'joint and concerted action by both the unions and the political party (the mass strike and the armed uprising . . .)'.[85] The other function was prior in time, being part of the process by which the revolution would be brought about. It was an educational function; and Lenin place 'the education of the proletariat' first when he was discussing the things that should be emphasized by party members undertaking 'revolutionary work in the trade unions'.[86] In other words, for Lenin, as for Marx and Engels, the trade union provided the organizational context within which workers would be raised to the level of revolutionary class-consciousness. *He*, however, thought of this purpose (raising workers' consciousness) as dependent upon the informed guidance of an élite party of professional revolutionaries for whom unions provided an important channel of communication with the working class. *They*, it seems, thought of the same purpose as dependent primarily upon the experience which trade union struggle directly provided for workers. Either way, the moulding of minds was a crucial issue; and trade unions were the vehicle.

[83] Quoted, in Bricianer, *Pannekoek and the Workers' Councils*, p.102.
[84] Lenin, *Collected Works*, v. 457.
[85] Ibid. xiii. 461.
[86] Ibid. 167.

5

THE ORGANICISTS

Trade Unions as Moral Forces

An organicist conception of society assumes that societies exhibit the kind of interdependence of parts characterizing literal organisms. On this view, the individuals, groups, classes comprising a society are defined, above all, by their functional contribution to its survival and well-being, in which all have a common interest. This fundamental and, above all, *morally* superior harmony of interests (so the line of thought runs) imposes on individuals and groupings an overriding obligation to prefer the common interest to their sectional interest in the event of a clash between them. To organicists, in other words, social analyses which focus exclusively or mainly on conflicting interests are wrong, because they neglect both the basic harmony and the responsibility of the parts to the whole that properly characterize human society. The task of right social theory is to rectify this distortion; and the task of right social action is to ensure that co-operation, rather than conflict, governs relations between the various groupings comprising a society.

In the specific case of industrial relations, the organicist thus assumes that any conflicting interests which an employer and his workers might appear to have are of far less consequence than the interests they undoubtedly do have in common—first, as 'partners' in industry and, second, as members of the larger society. Typical policy themes emerging from this approach are co-operative ventures, workers' participation (not control), co-partnership, profit-sharing, and, in general, a pronounced inclination to favour corporatist arrangements. The justification of these policies tends to be presented less in terms of the interests of the workers and employers directly affected, than in terms of the good of society at large.

Pushed to its limit, the organicist approach ends in the apotheosis of the state. The state is seen as in some sense embodying, in their totality, the interests of the whole society. The conclusion follows that the state is properly invested with final and exclusive authority for ensuring that the basic harmony of social interests is matched by an equivalent harmony of social action. This extreme version of organicism is considered, as a variant of authoritarianism, in Chapter 6.

The present chapter is concerned with versions of organicism which fall short of the authoritarian extreme in that they attach moral value to autonomous action on the part of associations other than the state. They are considered under the headings of Christian Socialism, Social Catholicism, and Conservatism. The writers dealt with are not necessarily the most prominent or the most representative of the broad sweep of opinion in their general grouping: attention paid to trade unionism was the main criterion of selection.

CHRISTIAN SOCIALISM

Christian Socialism has been fairly described as a 'difficult and ambiguous' notion.[1] As used here, it loosely covers a variety of lines of thought associated mainly with Protestant thinkers, not all of whom would have accepted the 'socialist' label. It was first used to identify a movement in mid-nineteenth-century Britain (though that movement itself was said to have a French inspiration).[2] It reached a peak of influence in both Britain and America in the latter years of the century; and there was a twentieth-century revival on both sides of the Atlantic. In the United States its less radical forms have been known as the Social Gospel movement and Social Protestantism.[3]

The fledgling movement figures in *The Communist Manifesto*: 'Christian Socialism is but the holy water with which the priest consecrates the heart-burnings of the aristocrat'.[4] Its founding fathers, notably F. D. Maurice and Charles Kingsley, were in fact associated with the Established Church. Nevertheless, like Christian Socialists of all varieties since, they were primarily concerned with reforms designed to improve the conditions, in the broadest sense, of working-class life. At the same time, while they

1 Jones, *The Christian Socialist Revival*, p. 453.
2 See Cole, *A History of Socialist Thought*, i. 291–2, 301.
3 Derber, *The American Idea of Industrial Democracy*, pp. 77, 151.
4 In McLellan, *Karl Marx: Selected Writings*, p. 239.

did not reject the state, the early Christian Socialists were 'as far as anyone could be from accepting State Socialism'.[5] Their emphasis was on voluntary action, and especially with an eye to 'the association of workers in the control of and the responsibility for production'.[6] For Maurice and Kingsley, what that meant, above all, was workers' co-operatives formed without state intervention, though with its ultimate protection. They also encouraged trade unions.

Later Christian Socialist groupings in Britain had strong links with the state-oriented Independent Labour party and Fabian Society. But it was Guild Socialism that most seductively tapped the resonating element in Christian Socialist thought, the worker-co-operative principle. S. G. Hobson himself was a Christian Socialist (one sign of it was his stress on society as 'a vast living organism');[7] but once G. D. H. Cole established himself as Guild Socialism's leading light, it largely lost its 'stained-glass character'.[8] There was, too, a Guild Socialist element in the writings of another, more famous Christian Socialist, R. H. Tawney.

Tawney argued that, 'subject to rigorous public supervision', those doing the actual work in an industry ('from organizer and scientist to labourer') should be responsible, through their 'professional organizations', for its conduct.[9] In other words, employees had a right 'to take part through their representatives in determining . . . an industry's policy . . . on matters which concern its prosperity and [their] own'.[10] This did not commit Tawney to the syndicalist road. He was thinking of workers' participation rather than workers' control. For he did not believe it was necessary completely to eliminate private ownership. He proposed, short of dispossession, a 'policy of attenuation'[11]—by which he meant circumscribing the rights of shareholders. He also made more of the unpleasant side of the responsibilities which workers' participation, as he viewed it, would thrust on to trade unions.[12] In general, however, Tawney showed little interest in the analysis of trade

[5] James, *The Christian in Politics*, p. 110.
[6] Greenslade, *The Church and the Social Order*, p. 116.
[7] Hobson, *National Guilds and the State*, p. 148.
[8] Jones, *The Christian Socialist Revival*, p. 292.
[9] Tawney, *The Acquisitive Society*, pp. 92, 176.
[10] Tawney, *Equality*, p. 267.
[11] Tawney, *The Acquisitive Society*, p. 100.
[12] See ibid. 152, 159.

unionism as an institution. Indeed, so far as the specific issue of trade union purpose is concerned, there has been only one strongly focused expression of a Christian Socialist position; and that came from the other side of the Atlantic.

Richard T. Ely published *The Labor Movement in America* in 1886. Like many others, he was moved by a horror of the social consequences of early industrialism, and a belief that industrialism, nevertheless, presented mankind with unprecedented opportunities which unfettered capitalism was incapable of realizing. He did not think of himself as a socialist, and scythed at 'the evils of socialism, nihilism, and anarchism'.[13] He did, however, admit that socialism at least had one redeeming feature, in that it aimed at 'a closer integration of social factors'.[14] Moreover, he leant towards conventional state socialism in a more specific sense. It was 'largely through the State' that the reforming power in society had to be brought into play.[15] He regretted the prevalence of antagonism towards the state—'the low view . . . too frequently taken of its nature'.[16] This had to change if it was to carry out its proper function. 'The Christian', he wrote, 'ought not to view civil authority in any other light than a delegated responsibility from the Almighty'; and the state's 'beneficent nature' ought to be made clear to workers in particular.[17]

On the other hand, important as the state was, it was for Ely only one of 'four chief agencies through which we must work' in the cause of social reform: 'These are the *labor organization*, the school, the State, and the Church'.[18] And the trade union ('labor organization') was as essential as the others. 'The organization of labor . . . is an indispensable condition of the improvement of the masses.'[19]

SOCIAL CATHOLICISM

The Catholic counterpart of Christian Socialism emerged in some strength during the latter half of the nineteenth century in Europe, especially in France, Germany, Austria, and Switzerland. It, too, rejected state socialism as a solution to the social consequences of

[13] Ely, *The Labor Movement in America*, p. 324.
[14] Ibid. 288.
[15] Ibid. 325.
[16] Ibid.
[17] Ibid. 326, 329.
[18] Ibid. 324. Emphasis added.
[19] Ibid. 323.

industrialism and, while conceding a major role to the state, emphasized non-governmental associations as an important source of reform. Social Catholicism, as it was generally known, from its beginnings reached back intellectually to the Middle Ages, to Thomas Aquinas and to the medieval guild system. As a result, it was propagating a modernized version of the guild idea long before Guild Socialism—which itself, by the 1920s, had a conservative Social Catholic wing. In contrast to Guild Socialism, however, mainstream Social Catholics not only believed in private enterprise but, for a time, tended to think of trade union organization literally along the lines of medieval guilds. Hence the promotion of 'mixed unions' which included owner-employers as well as their employees.[20] The belief in mixed unions was to wane (although they were legally fostered up to the 1970s in Spain and Portugal), but Social Catholicism is still characterized by the underlying assumption that trade unions, properly conceived, have a capacity to reconcile employers and employees.

Ideas floated under the banner of Social Catholicism, by such men as Bishop von Ketteler and Count Albert de Mun, first received the ultimate stamp of approval in 1891 when Pope Leo XIII published the encyclical letter, *Rerum Novarum* (otherwise most commonly known as 'On the Condition of Labour'). The 'condition of the working classes', the Pope said, 'is the pressing question of the hour'.[21] He condemned economic liberalism ('unchecked competition')[22] as its primary cause, and rejected socialism as its solution. The true solution depended upon the fact that, contrary to the socialist argument, capital and labour were not 'naturally hostile':

> Just as the symmetry of the human frame is the result of the suitable arrangement of the different parts of the body, so in a State is it ordained by nature that these two classes should dwell in harmony and agreement, so as to maintain the balance of the body politic. Each needs the other.[23]

The true solution also depended upon the combined efforts of three agencies above all—the Church, the state, and 'workingmen's unions'.[24]

[20] Moon, *The Labor Problem and the Social Catholic Movement in France*, p. 392.

[21] *Rerum Novarum* in Gilson, *The Church Speaks to the Modern World*, p. 238.

[22] Ibid. 206. [23] Ibid. 214. [24] Ibid. 231.

The state, according to *Rerum Novarum*, has a protective role in relation to which the poor would properly claim 'especial consideration' because they lacked the independent resources of the rich: 'it is for this reason that wage-earners . . . should be specially cared for and protected by the government'.[25] The state, accordingly, should concern itself (preferably, to avoid 'undue interference', by way of special 'societies or boards') with the fairness of wages, working hours and industrial health and safety arrangements.[26] In addition, the state should not prohibit trade unions. It was 'the natural right of man' to enter 'private societies' of this kind.[27] The State should 'watch over' them; but it 'should not thrust itself into their peculiar concerns and their organization'.[28] The Pope, remarking that many existing trade unions had undesirable features, expressed a preference for unions consisting solely of 'Christian working men'.[29] On the other hand, his words did not rule out what has elsewhere been described as 'the neutral type [of trade union] if they pursue legitimate purposes'.[30]

Rerum Novarum has been followed by three other papal encyclicals professing to elaborate its principles. The first of these, published on its fortieth anniversary in 1931, is said to have been influenced by an American, Father John A. Ryan, who had in turn drawn inspiration from the work of Richard Ely, under whom he studied. Ryan's main point in an early article on 'moral aspects' of trade unions was that they were 'a necessary evil', in the sense that their 'evil effects' were, 'on the whole, morally outweighed by [their] good effects'.[31] In *Quadragesimo Anno* ('On Reconstructing the Social Order') in 1931, Pius XI defended trade unions in a different way by voicing specific, if guarded, criticism of the controls imposed on Italian trade unionism by the current Fascist regime.[32]

Excessive use of state power, and not only in relation to trade unions, was also among the concerns of *Mater et Magistra* ('On Christianity and Social Progress') published in 1961 by Pope John XXIII. For his part, Pope John Paul II, in *Laborem Exercens* ('On Human Work') published in 1981, displayed more direct interest in

[25] *Rerum Novarum* in Gilson, *The Church Speaks to the Modern World*, pp. 225–6. [26] Ibid. 230. [27] Ibid. 232–3. [28] Ibid. 235.

[29] Ibid. 234. [30] Moody, 'Leo XIII and the Social Crisis', pp. 81–2.

[31] Ryan, 'Moral Aspects of Labor Unions', pp. 724, 728.

[32] See *Quadragesimo Anno* in Treacy and Gibbons, *Seven Great Encyclicals*, pp. 150–1.

trade unions. He described them as 'an indispensable *element of social life*' in modern societies.[33] He emphasized, too, that trade unions were essentially about harmony, not conflict, despite their involvement in struggle.

However, this struggle . . . *is not a struggle 'against' others*. Even if in controversial questions the struggle takes on a character of opposition towards others, this is because it aims at the good of social justice, not for the sake of 'struggle' or in order to eliminate the opponent. It is characteristic of work that it first and foremost unites people. In this consists its social power: the power to build a community. In the final analysis, both those who work and those who manage the means of production or who own them must in some way be united in this community. *In the light of this fundamental structure* of all work . . . it is clear that, even if it is because of their work needs that people unite to secure their rights, their union remains a constructive factor of *social order* and *solidarity*.[34]

CONSERVATISM

Writers designated by others as Conservative span a long continuum. Peter Viereck underscored this when he compiled an anthology on the following assumption: 'no "pure" conservatives ever exist; all are diluted in varying degrees with either authoritarianism or liberalism—Maistre and Burke being the respective prototypes of these two dilutions'.[35] And there is plenty of room for confusion at the margins. Thus F. A. Hayek and Milton Friedman have been described as 'liberal-conservatives',[36] although each had specifically rejected the Conservative label: Friedman preferred to be thought of as an adherent of 'liberalism in its original sense', while Hayek, more wary of the modern American usage of 'liberal', preferred 'an unrepentant Old Whig'.[37]

The tension in Conservative thought, between the value placed respectively on political authority and political freedom, is more pronounced than in the case of either Christian Socialism or Social Catholicism. Another distinctive feature is that a religious

[33] John Paul II, *Laborem Exercens*, p. 73.
[34] Ibid. 73–4.
[35] Viereck, *Conservatism*, p. 6.
[36] O'Sullivan, *Conservatism*, p. 139.
[37] See Friedman, *Capitalism and Freedom*, p. 6; Hayek, *The Constitution of Liberty*, p. 409.

under-pinning cannot be taken for granted: 'while there is a connection between conservative and religious feeling, it is now difficult to argue for their identity'.[38] Nevertheless, like other organicists, Conservatives have commonly assumed that the interests which workers and their employers share are of greater moral and social significance than the interests they do not share. Thus there has been, among them, the same perception of industry as essentially a joint enterprise. Sometimes, this has been associated with an attribution of value to the role of trade unionism. But, more often than not, Conservative writers have found in trade unions a source of unease rather than hope. Thomas Carlyle is an early case in point.

Carlyle faced the task of adapting the Conservatism of Edmund Burke, with its assumption of an existing organic society, to the society created by early nineteenth-century industrialism. Looking back to medieval models, he judged contemporary British society (in 1829) to be mechanical in character, rather than organic: 'We figure society as a "Machine" '.[39] The central characteristic of this 'disorganic . . . and hell-ridden world' was that worker-employer relations in industry were based on money alone.[40] Yet it was 'not by Self-interest, but by Loyalty, that men are governed or governable';[41] and money could not buy loyalty, the cement of organic relationships. The solution, Carlyle thought, lay in the creation of a 'Chivalry of Work', a 'Chivalry of Labour', involving 'a blessed loyalty of Governor and Governed'.[42] Military chivalry, deriving from feudal times, provided the model. Just as 'a Fighting World had to be regimented, chivalried', so did 'a Working World', with 'laws and fixed rules which follow out of that,—far nobler than any Chivalry of Fighting was'.[43] Two things, above all, were required to achieve this. One was permanent employment: 'Permanence, persistence is the first condition of all fruitfulness in the ways of men.'[44] The other was organization; and the ' "Organising of Labour" is . . . the Problem of the whole Future'.[45]

[38] Scruton, *The Meaning of Conservatism*, p. 170.
[39] Carlyle, *Scottish and other Miscellanies*, p. 239.
[40] Carlyle, *Past and Present*, p. 391.
[41] Carlyle, *Scottish and other Miscellanies*, p. 221.
[42] Carlyle, *Past and Present*, p. 392.
[43] Ibid. pp. 370–1.
[44] Ibid. 376.
[45] Ibid. 351–2.

Carlyle, however, displayed not the slightest inclination to entrust the organizing of labour (or, indeed, anything else) to trade unions. He wrote of the 'Trades Union . . . with assassin pistol in its hand'.[46] He wrote of its membership: 'O mutinous Trades-Unionist, gin-vanquished, undeliverable'.[47] And he wrote of the company that trade unions kept in his mind: 'Work cannot continue. Trades' Strikes, Trades' Unions . . . mutiny . . . rage and desperate revolt . . . will go on their way.'[48] Trade unions, it followed, had no part in the chivalrous organization of labour. That depended on industrial employers. They alone could provide the 'Industrial Hero', the 'Industrial noble', whose function it would be 'to recivilise . . . the world of Industry'.[49] The lengths to which such a man might go could only be guessed at.

A question arises here: Whether, in some ulterior, perhaps some not far-distant stage of this 'Chivalry of Labour', your Master-Worker may not find it possible, and needful, to grant his Workers permanent interest in his enterprise and theirs? *So that it become, in practical result, what in essential fact and justice it ever is, a joint enterprise;* all men from the Chief master down the lowest Overseer and Operative, economically as well as loyally concerned for it?[50]

There is in Carlyle's notion of industry as a joint enterprise, 'in practical result', more than a germ of the co-partnership proposals of a later age.

Lord Salisbury later echoed Carlyle's attitude to trade unions when he wrote bitingly of their 'strong, steady, deadly grip', of 'the stupid barbarism of their economical creed', and of 'their terrible organization'.[51] Viewing them at a time when they were more firmly established, however, Salisbury (in contrast to Carlyle) was repelled precisely by their accomplishments as organizers. For this meant that 'workmen' were able to 'act *en masse* with a success which no class or order of men not bound together by religious ties has ever succeeded in attaining to before'.[52]

This, it should be said, did not mean that the Conservative governments Salisbury led (or those of Disraeli beforehand and

[46] Carlyle, *Scottish and other Miscellanies*, p. 325.
[47] Carlyle, *Past and Present*, p. 394. [48] Ibid. 399.
[49] Carlyle, *Scottish and other Miscellanies*, p. 324.
[50] Carlyle, *Past and Present*, p. 382. Emphasis added.
[51] In Smith, *Lord Salisbury on Politics*, pp. 143, 176.
[52] Ibid. 212.

Balfour afterwards) actually treated trade unions as beyond the pale. Each, on the contrary, moved to deal with them in some measure:[53] Salisbury himself was, in fact, the first Conservative prime minister in British history to receive a deputation from the Trades Union Congress.[54] But, as distinct from political practice, there seems to have been no significant *published* change of tune in the Conservative camp until 1912. And then it amounted to a virtual volte-face. Lord Hugh Cecil flatly affirmed the social value of trade unions. His central point was that 'the voluntary action of trade unions has served the working-class better than any exertion of the powers of the State could have done'.[55]

By the 1930s, Cecil's acceptance of the value of trade unions was being translated into an acceptance of them as participants in the administration of industry. This occurred in Harold Macmillan's proposal for a reorganization of industry which involved, among other things, an elaborate consultative apparatus designed to bring 'labour more and more into active partnership' with management.[56] Trade union participation was central to this structure, which Macmillan thought 'would provide a counterpart, in the economic sphere, to the political self-government . . . achieved through the extension of the franchise'.[57] Later, he was less coy, and spoke of it simply as 'industrial democracy'.[58]

Macmillan, it must be said, was the voice of a radical ginger group within the British Conservative party. Contemporaries closer to mainstream Conservatism have preferred to dwell rather more on the failings of trade unionism. Quintin Hogg (later Lord Hailsham) echoed Macmillan when he affirmed that 'trade unions, free, independent, and powerful, have an honourable and authoritative part to play in industrial organization'.[59] But then, unlike Macmillan, he went on to emphasize their shortcomings. In the later versions of his book,[60] moreover, the affirmation was dropped and the exposition of union deficiencies amplified. Similarly, while T. E. Utley conceived of industry as a partnership and advocated a

[53] See Martin, *TUC*, pp. 29–33, 36–7, 62–3, 93–5.
[54] Ibid. 93. [55] Cecil, *Conservatism*, p. 190.
[56] Macmillan, *The Middle Way*, p. 217.
[57] Macmillan, *Reconstruction*, p. 122.
[58] Macmillan, *The Middle Way*, p. 217.
[59] Hogg, *The Case for Conservatism*, p. 126.
[60] Hailsham, *The Conservative Case* (1959); Hailsham, *The Dilemma of Democracy* (1978).

form of worker participation, he saw no enduring trade union role in the joint enterprise. On the contrary, the realization of full worker-management co-operation would 'remove the *raison d'être* of trade unions, which exist to bargain not to co-operate'.[61]

It was left to Frank Tannenbaum, on the other side of the Atlantic, to combine (and enlarge) the Cecil–Macmillan valuation of unionism with Carlyle's vision of an organic industrial society, by nominating the trade union for the regenerating role that Carlyle had allotted to employers alone. There was no doubt, for Tannenbaum, about the essentially organic nature of the linkage, not merely between workers and employers, but specifically between trade unions and management: 'the quarrel between the labor union and management has always been a family quarrel. They developed together, were interdependent, and expressed different aspects of the same institution.'[62] The 'continuing cooperation' between them had been obscured by strikes and other dramatic forms of industrial conflict.[63] The task was to bring the underlying reality to the surface: 'in some way the corporation and its labor force must become one corporate group and cease to be a house divided and seemingly at war'.[64] That, in any case, was precisely the direction in which history was moving—towards a time when the 'corporation and the union will ultimately merge in common ownership'.[65] And trade unions had the major part to play in the process.

The significance of trade unionism was a function of its nature, as 'an organic growth in modern society'.[66] In other words, like the guilds of medieval and ancient societies, trade unionism reflected 'a sense of identity among men laboring at a common task'.[67] By the same token, it embodied as well 'a revulsion against social atomization . . . and [against] the divorce of owner and worker from their historical function as moral agents in industry'.[68] Tannenbaum thought that the trade union was in this sense 'the only *true* society that industrialism has fostered'.[69]

The singular character of trade unionism had one consequence of immense historical importance. For it entailed the role, in relation to workers, of 'creating a sense of identity with industry'.[70]

[61] Utley, *Essays in Conservatism*, p. 30.
[62] Tannenbaum, *The True Society*, p. 82.
[63] Ibid. 177. [64] Ibid. 168–9. [65] Ibid. 199. [66] Ibid. 99.
[67] Ibid. 14. [68] Ibid. 105. [69] Ibid. 198. [70] Ibid. 177.

Trade-unionism is the conservative movement of our time. It is the counterrevolution ... In tinkering with the little things—hours, wages, shop conditions, and security in the job—the trade-union is ... rebuilding our industrial society upon a different basis from that envisioned by the philosophers, economists, and social revolutionaries of the eighteenth and nineteenth centuries.[71]

That 'different basis' amounts, in spirit, to what Carlyle had described as a 'Chivalry of Labour'; and Tannenbaum's 'identity with industry' is synonymous with Carlyle's 'Loyalty'.

WHOSE INTERESTS?

The responsibility of the part to the whole is inseparable from the idea that society is an organism. For the trade union that means a responsibility which extends beyond the membership, beyond the class, to society at large. Thus Richard Ely looked on trade unions as one of the agencies 'through which we must work for the amelioration of the laboring class, as well as of *all* classes of society'.[72] Similarly, Pope John Paul II described the 'specific role' of trade unions as being 'to secure the just rights of workers within the framework of the common good of the whole of society'.[73] The same position is implicit in Conservative writings.

One point, however, needs to be emphasized by way of contrast with the organicists' authoritarian counterparts considered in the next chapter. In the case of all three organicist variants (with the exception of some early writers), there is a general recognition of an area within which a trade union is properly free to act as if its only responsibility were to its members. That area relates to the negotiation (including, in principle, use of the strike weapon) of wages and working conditions. There are, of course, boundaries to this freedom. But the critical point is the organicists' acceptance, however notional, of an area in which the state lacks the moral authority to intervene unilaterally in order to enforce a perceived obligation to society at large. There can be no such area for the authoritarian.

INTERESTS IN WHAT?

Organicist theories of trade union purpose eschew revolutionary aims. But all, in some measure, proffer reform or improvement as a

[71] Tannenbaum, *The True Society*, p. 3.
[72] Ely, *The Labor Movement in America*, p. 324. Emphasis added.
[73] John Paul II, *Laborem Exercens*, p. 75.

goal. Typically, too, they emphasize the connection between social reforms, on the one hand, and ethical values on the other.

Richard Ely

For Ely, one aspect of social reform had to do with improving workers' industrial and economic conditions. He acknowledged this, as a purpose of trade unions, in a chapter entitled 'The Economic Value of Labor Organizations'.[74] He had, however, a different and broader purpose in mind when he wrote of American trade unions 'playing a role in the history of civilization, the importance of which can be scarcely overestimated'.[75] That purpose was nothing less than the moral improvement of the working class and, through it, society at large. The function this purpose entailed was an educational function. And it was their educational function, above all, which accounted for the enormous importance he attached to trade unions: 'they are among the foremost of our educational agencies . . . in their influence upon the culture of the masses. They reach and elevate large classes, mentally, morally, and spiritually, who can be moved in no other manner'.[76] As this passage suggests, Ely was thinking of 'education' in the broadest sense. 'I mean . . . the entire development of a man in all his relations, social, individual, religious, ethical, and political.'[77]

Thus trade unions were schools—but never, even in their use of strikes, schools of war as they were for Engels. Rather, they were 'schools of political science' in which members discussed political issues.[78] Participation in their affairs otherwise provided an 'excellent training in a practical school of politics'[79] (a point later

[74] Ely, *The Labor Movement in America*, pp. 92–119.

[75] Ibid. 120.

[76] Ibid. R. H. Tawney seems never to have thought of trade unions in this role. Possibly he was under the spell of Sidney and Beatrice Webb (they had not thought of it either) for whose writings on trade unions he plainly, and justifiably, had great admiration (see Tawney, *The Attack*, pp. 110–13). At any rate, he never got beyond a view of trade unions as having basically a 'dual character as a body of professional associations and an organ of agitation' to which, 'half-unconsciously', a third role involving a responsibility for 'more general issues of economic strategy' had been lately added (see Tawney, *The Radical Tradition*, p. 88). This is worth particular note in view of his intimate involvement with the Workers' Educational Association from its beginnings; and his belief in the importance of educational opportunities to workers (see e.g. Tawney, *The Radical Tradition*, pp. 88, 93).

[77] Ely, *The Labor Movement in America*, p. 120.

[78] Ibid. 124. [79] Ibid. 127.

echoed by Archbishop Temple, a celebrated English Christian Socialist who wrote of 'the schooling in democratic habits provided . . . by the Trade Unions').[80] As part of this practical training, the unions were also 'schools of oratory'.[81] Less predictably, they were 'schools in which true politeness and even grace of manner are taught'—a proposition which Ely supported with a footnoted query: 'How many rich men's clubs . . . impose fines for profanity?'[82]

Nor did the educational function of trade unions stop there. Apart from their role as 'perhaps the chief power . . . making for temperance', they were the means by which workers 'are learning discipline, self-restraint, and the methods of united action'.[83] Similarly, for reasons which Ely explained at some length, unions 'educate the laborers to prudence in marriage'.[84] And, of 'still wider ethical significance', as he put it, they were not only capable of introducing 'a higher tone into our political life', but they comprised 'the strongest force outside the Christian Church making for the practical recognition of human brotherhood'—and that, he believed, had a bearing on world peace.[85]

The Popes and Father Ryan

In *Rerum Novarum* Leo XIII, if more briefly and with more caution than Ely, set out a view of trade union purpose which similarly embraced both the economic and the ethical. The trade unions' 'immediate purpose is the private advantage' of their members, and that was primarily a matter of economics.[86] But there was a larger and less immediate purpose as well. Proper trade unions were under an obligation to 'pay special and chief attention to the duties of religion and morality'.[87] The purpose of attending to religion and morality clearly implied an educational function.

Father Ryan, for his part, was to describe the purpose of trade unions simply as 'higher wages and other improvements in [workers'] economic condition'.[88] Subsequently, he expanded that to include 'labor participation in management', profit-sharing and co-operative ownership.[89] He did not follow *Rerum Novarum* to

[80] Temple, *Christianity and Social Order*, p. 73.
[81] Ely, *The Labor Movement in America*, p. 127.
[82] Ibid. 135. [83] Ibid. 130, 136. [84] Ibid. 117. [85] Ibid. 137–8.
[86] *Rerum Novarum* in Gilson, *The Church Speaks to the Modern World*, p. 232.
[87] Ibid. 236. [88] Hillquit and Ryan, *Socialism*, p. 135.
[89] Ryan, *Declining Liberty*, pp. 230–5.

specify religion and morals, but they may have been implied in a passing reference to 'education' as a desirable union function.[90] However that may be, Ryan (while acknowledging the unions' primary concern with their collective bargaining function) later emphasized education as a function which, with workers' participation in mind, was uniquely appropriate to the trade union. 'That is education of the workers in self-respect, in the conviction that they are fitted . . . to take a gradually and indefinitely increasing share in the management . . . of industry.'[91]

Quadragesimo Anno took up Ryan's advocacy of worker participation, co-operative ownership, and profit-sharing, urging arrangements that would enable workers to 'become sharers in the ownership or management, or else participate in some way in the profits'.[92] Pius XI also softened the implications of *Rerum Novarum* by explicitly authorizing what Leo XIII had only tacitly allowed. Catholic membership of 'neutral trade unions' was permissible provided, among other things, that there were parallel 'associations which aim at giving their members a thorough religious and moral training'.[93] The hope was that these members would in turn influence the moral climate of the unions to which they belonged. The point was raised again in *Mater et Magistra* by way of the accolade which John XXIII bestowed on those 'beloved sons of ours' working within trade unions 'to vindicate the rights of workingmen and to improve their lot *and conduct*'.[94] He, too, stressed worker participation in management, but extended the principle to include support for trade union participation in national and international governmental institutions concerned with 'various sectors of economic life'.[95]

The full gamut of Ryan's proposals (worker participation, joint ownership, and profit-sharing) again received papal endorsement in *Laborem Exercens*[96]—along with co-partnership. The encyclical also specified the general trade union purpose enclosing these particular aims: 'to defend the existential interests of workers in all sectors in which their rights are concerned'.[97] Then the moral and

[90] Ryan in *Catholic Encyclopaedia*, p. 728.
[91] Ryan, *Declining Liberty*, p. 221.
[92] Treacy and Gibbons, *Seven Great Encyclicals*, p. 144.
[93] Ibid. 134.
[94] Ibid. 240. Emphasis added.
[95] Ibid.
[96] John Paul II, *Laborem Exercens*, p. 53.
[97] Ibid. 73.

functional implications of this purpose were spelt out in a way which carefully stressed the educational aspect.

Speaking of the protection of just rights of workers . . . before all else, we must keep in mind that which conditions the specific dignity of the subject of the work. The activity of union organizations opens up many possibilities in this respect, including their *efforts to instruct and educate* the workers and to *foster their self-education*. Praise is due to the work of the schools, what are known as workers' or people's universities . . . It is always to be hoped that, thanks to the work of their unions, workers will not only *have* more, but above all *be* more: in other words, that they will realize their humanity more fully in every respect.[98]

Tannenbaum and Others

On the side of the Conservatives, Lord Hugh Cecil also placed weight on an educational function. In contrast to John Paul II, however, he regarded it as a matter of experience rather than formal training.

Workmen combining together in a trade union to get better wages or shorter hours obtain not only the wages or the hours for which they strive, but a most valuable social and political education by the way. They have to learn to work with one another; they have to learn to respect public opinion; they have to learn to be reasonably regardful of the interests of other persons.[99]

As this passage indicates, Cecil not only accepted the goal of advancing workers' material interests, but imputed to trade unions an implicit moral purpose as well. For Harold Macmillan, on the other hand, worker participation in management was the focus (and he was sympathetic, in principle, to co-partnership and profit-sharing as well).[100] He regarded the 'recognition and encouragement of the workers' Trades Unions' as vital to the general interest in forging an 'active partnership' between management and labour.[101] In certain circumstances, he was even prepared to accept the legal enforcement of union membership as a means of cementing such a partnership.[102]

98 John Paul II, *Laborem Exercens*, pp. 75–6. Emphasis as in original.
99 Cecil, *Conservatism*, pp. 191–2.
100 See Macmillan, *Reconstruction*, p. 121.
101 Macmillan, *The Middle Way*, p. 217.
102 See ibid. 216.

Trade union purpose for Frank Tannenbaum, however, was a much broader conception (though, in the 1950s, not quite as broad as it had been thirty years earlier when he had espoused a form of syndicalism).[103] For a start, there was the material purpose, which he explained in terms reminiscent of Selig Perlman: 'The need for economic security is implicit in almost every trade-union demand.'[104] Yet, while trade unions 'may talk the language of the market and be obsessed by economic objectives',[105] there were other layers of purpose. Thus there was the goal of 'increased participation in management'; and that was in fact involved in each dispute on specific economic issues.[106] 'But', probing deeper, 'the underlying theme is the drive for moral status within the industry.'[107] The trade union, in other words, 'has a moral purpose'.[108] And that moral purpose was the creation of a sense of identity between workers and their work.

> The union, if it is to survive, must maintain the fealty of its members, and can do that only by giving them a sense of dignity and standing, not only within the union, but also within the industry. Such a sense of dignity can only be had if the workers have a concern for all the issues and difficulties of the enterprise.[109]

One way in which American trade unions were implanting a sense of dignity, Tannenbaum maintained, was through their growing practice of investing funds in corporations employing their members. Such investment amounted to 'the re-establishment of a proprietary interest by the workers in the industries from which they draw their living'.[110] It was a policy that he thought was destined to bring about a fundamental alteration in the attitudes of employees.

> For the worker, at least, the day of fluidity, impersonality, and irresponsibility is drawing to a close. If he is going to have a pecuniary interest, he will also have to assume the moral responsibilities that a pecuniary relationship has always involved: responsibility for the property he owns, the work he does, and the quality of what he produces for the rest of the community.[111]

103 See Tannenbaum, *The Labor Movement.*
104 Tannenbaum, *The True Society*, p. 115.
105 Ibid. 163. 106 Ibid. 107 Ibid.
108 Ibid. 164. 109 Ibid. 163.
110 Ibid. 189. 111 Ibid.

This underlying moral purpose of trade unions was what truly defined their character as institutions instinctively designed to recapture a lost past.

> Trade-unionism is a repudiation of Marxism because its ends are moral rather than economic. It is a social and ethical system, not merely an economic one. It is concerned with the whole man. Its ends are the 'good life'. The values implicit in trade-unionism are those of an older day . . . It is an unwitting effort to return to values derived from the past . . . [in which] man had found his human dignity.[112]

And the strength of trade unionism lay precisely in its ability to satisfy 'the human craving for moral status in a recognizable society'.[113]

Given the overriding moral purpose of the trade union, it followed that the union's central function was an educational function. As Tannenbaum put it in an early work, trade unionism educated members in the 'very process' of organizing them; and what it provided was 'chiefly an education in morals: in responsibility, in character and in initiative'.[114] Carlyle would have been thunderstruck at the thought that his 'Industrial Hero', creator of the 'Chivalry of Labour', was in fact to come from the side of labour itself.

It would be grotesque, however, to represent Tannenbaum's views (and, indeed, Cecil's or Macmillan's) as typical Conservative interpretations of twentieth-century trade unionism. Each of them offered a perspective that, while identifiably within the Conservative tradition, was nevertheless outside the mainstream of that tradition. Carlyle, certainly, would have been much more comfortable with the less atypical interpretation of two other American writers. 'By and large the union's societal character has disappeared. Its moral impulse and democratic spirit have largely been dissipated . . . As a fellowship or society, it has grown old and seems about to die.'[115] Or, as a Joseph Heller character had it: 'The labor movement [has] come to its end in garbage strikes and gigantic pension funds.'[116] To be sure, Carlyle (not to mention

[112] Tannenbaum, *The True Society*, pp. 10–11.
[113] Ibid. 13.
[114] Tannenbaum, *The Labor Movement*, p. 238.
[115] Hill and Stuermann, *Organized Labor: A Philosophical Perspective*, p. 175.
[116] Heller, *Good as Gold*, p. 74.

Salisbury) would have hotly disputed the underlying assumption of these comments—that trade unions had once been something other than self-seeking and anti-social. On the other hand, he would unquestionably have accepted their implication that the central purpose of trade unions *ought* to be a moral purpose of the kind discerned by Tannenbaum.

6

THE AUTHORITARIANS

Trade Unions as State Instruments

The authoritarian is characterized by an overwhelming emphasis on the role of the state or of a ruling political party. Trade unions are allotted a decisively subordinate role. In some cases their subordination is justified with reference to an organic conception of society. In all cases, it is associated with the goal of maximizing industrial production.

Authoritarians comprise a motley group in conventional political terms. As represented below, they range from Mussolini and Hitler on the right, to Lenin and Stalin on the left; and in between, more or less, Juan Perón of Argentina, Tom Mboya of Kenya, and Sidney and Beatrice Webb of England. Each of these authors is identified with a specific ideology or political movement, and that is acknowledged in the section headings that follow. On the other hand, the focus of this chapter is the authors themselves, rather than the movements with which they were associated. This means, for example, that while the Webbs are fairly identified with Fabian socialism, there is no suggestion that all Fabian socialists necessarily accept the Webbs' conception of trade union purpose under socialism, as presented below.

ITALIAN FASCISM

Benito Mussolini, founder of the National Fascist party and dictator of Italy from 1922 to 1943, employed the organicist analogy with particular frequency and zest. The Fascist state was to be 'strong and organic'.[1] The individual was to be accepted 'only in so far as his interests coincide with those of the State'.[2] And Mussolini's focus was precisely the state, not society or nation. For 'the nation is created by the State'.[3] Such a state, too, was

[1] Mussolini, *Fascism*, p. 29. [2] Ibid. 10.

[3] Mussolini in Oakeshott, *Social and Political Doctrines of Contemporary Europe*, p. 167.

necessarily all-enveloping: 'everything in the State, nothing against the State, nothing outside the State'.[4] In other words, 'nothing human or spiritual exists, much less has value, outside the State'.[5]

It followed that the achievement of the organicist ideal of social harmony depended ultimately on the state alone, because only the state was capable of transcending the mêlée of conflicting interests that otherwise disfigured society. Above all, it was specifically the Fascist state which most completely 'concentrates, controls, harmonises and tempers the interests of all social classes'.[6]

In the particular case of the organization of working-class interests, trade unionism used as 'a class weapon' was to be condemned as divisive.[7] But trade unionism incorporated within the unity of the Fascist state was a different matter. It had a justified place in the sun. The terms on which it did were stated clearly enough in the 1927 'Charter of Labour' (which, according to Mussolini, 'I made the Great Council approve').[8] 'Occupational or syndical organization is free; but only the juridically recognized syndicate which submits to the control of the State has the right to represent legally the entire category of . . . workers for which it is constituted.'[9]

GERMAN NATIONAL SOCIALISM

Adolf Hitler, unlike Mussolini, had still to win power when he set out his views on trade unions in *Mein Kampf* (*My Struggle*). German trade unionism, too, was historically stronger and more influential than its Italian counterpart. This may help explain why he was at pains to demolish the idea that trade unionism was 'essentially hostile to the Fatherland'.[10]

> If Trades Union action aims at improving the condition of a class which is one of the pillars of the nation, and succeeds in doing so, its action is not against the Fatherland or the State, but is 'national' in the truest sense of the word.[11]

But this honourable kind of trade unionism ('an instrument for defending the social rights of the employee') had to be sharply

[4] Mussolini, *Fascism*, p. 40.
[5] Mussolini in Oakeshott, *Social and Political Doctrines of Contemporary Europe*, p. 166.
[6] Mussolini, *Fascism*, p. 41.
[7] Ibid. 11.
[8] Mussolini, *My Autobiography*, p. 256.
[9] Oakeshott, *Social and Political Doctrines of Contemporary Europe*, p. 184.
[10] Hitler, *My Struggle*, p. 25.
[11] Ibid.

distinguished from the kind of unionism which was used 'as a party instrument in the political class war'.[12] The second kind was a perversion. In so far as German trade unionism had been twisted in that direction, two malevolent forces were responsible. One was Marxism. Acting through the German Social Democratic party, Marxism had largely converted the trade union movement into 'the battering ram for the class war'.[13] The other force, inevitably for Hitler, was the Jews, who were 'gradually assuming leadership' of the unions.[14] The combination was deadly. Marxism, in converting trade unions to its own purposes, created the instrument which was then used by the international Jewish conspiracy to destroy 'the economic basis of free and independent national States'.[15] Trade unionism, an institution of 'immense importance' which might otherwise have been 'the saving of the nation', had thus been forged into 'one of the most terrible instruments of intimidation'.[16]

The trade unionism of the future National Socialist state, Hitler predicted, would be different; and it was. The German Labour Front was 'a Nazi organization, its functionaries Nazis, its objectives Nazi objectives'.[17] And, as Hitler had also predicted, there was no room for any other kind of trade union. Robert Ley, head of the Labour Front, explained why in classically organicist terms: 'it was impossible to fit infected organs into the new social body'.[18]

ARGENTINE PERONISM

Colonel Juan Domingo Perón was president of Argentina from 1946 until his overthrow by the military in 1955, and again for the nine months before his death in 1974. During a two-year stint as Secretary of Labour and Social Welfare in an earlier military government, he had forged ties with a section of the trade unions and secured a breadth of popular support which enabled him to survive a palace revolt in 1945, and win his first presidential election the following year. His wife, Eva Duarte de Perón, controlled the Secretariat of Labour and the trade union movement during the first six years of his presidency.

[12] Hitler, *My Struggle*, p. 24. [13] Ibid. 26.
[14] Ibid. 128. [15] Ibid. 239. [16] Ibid. 25–6, 129.
[17] Schoenbaum, *Hitler's Social Revolution*, p. 90.
[18] Quoted in Brady, *The Spirit and Structure of German Fascism*, p. 120.

Perón often spoke in organicist terms, and on occasion used the quite different metaphor of the machine. In both cases, however, his point was the same. The whole was dependent on the parts, but greater than them: 'There is nothing superior to the interest of the whole.'[19] This principle, applied to society, meant that it was 'the duty of the State, having the superior interests of [all parts of society] in view, to coordinate them'.[20] The Perónist 'movement' ('not a political party') shared this perspective. It, too, did not stand for 'sectarian or party interests'; rather, 'we represent only the national interests'.[21] A fundamental objective of the movement was 'to end class war and substitute cooperation between capital and labor'.[22]

Trade union organization, as Perón put it, 'is the basis of our procedure'.[23] He incorporated the 'right freely to form trade unions' in the Argentine Constitution.[24] But, he warned, it was a right that had its dangers. In particular, trade unions might be diverted from legitimate purposes by the introduction of 'politics' into their affairs; and politics in trade unions generated disunity.[25] Disunity meant an 'inorganic working mass', which was easy prey for 'professional agitators in the pay of . . . foreign powers'.[26] Unified trade unions were therefore essential, and the state had a crucial role in this respect. In the first place, 'unions inspired by secondary political and ideological intentions should be put on one side by the law'.[27] Otherwise, the state needed to foster trade unions by 'protecting them . . . giving them a system of absolute security, and aiding them in their progress'.[28] Moreover, special attention had to be paid to their officials. They needed to be 'good leaders, who are selected';[29] it was crucial to exclude 'all extremists whose ideologies appear to us to be of . . . an exotic nature'.[30]

The great political significance that Perón attached to the trade union movement was underlined by the emphasis he placed on its centralized control.

The government needs . . . a large labour centre, as powerful as possible . . . in order to be able to fulfil the great destiny of this country . . . A

[19] Perón, *Peronist Doctrine*, p. 28.
[20] Ibid. 121. [21] Ibid. 186.
[22] Eva Duarte Perón, *Evita by Evita*, p. 83.
[23] Perón, *Peronist Doctrine*, p. 74. [24] Ibid. 339.
[25] Ibid. 351. [26] Ibid. 348, 423. [27] Ibid. 348.
[28] Ibid. 350. [29] Ibid. 352. [30] Ibid. 424.

powerful labour centre is the best guarantee for the government, which has no other power than the power of labour.[31]

AFRICAN SOCIALISM

The new states of tropical Africa, which emerged from the late 1950s, shared with other 'developing' nations both their aspirations for economic growth and the handicaps that typically impede development in poor countries. Many African leaders and intellectuals placed particular weight on what they assumed to be uniquely African aspects of the problems and their solutions. This emphasis underlay an influential school of thought which postulated a need 'to create an African personality and an African socialist system'.[32] The words are those of Tom Mboya, a major trade union official who became, and remained, a leading member of the government of independent Kenya until his murder in 1969.[33]

African trade unions in colonial times usually provided an important source of support for nationalist movements, a role which encouraged an assertive style. Mboya warned that, with the achievement of independence, the style had to change. 'Restraint', instead, was required of them—and not only because that was essential for economic growth but also, he added ominously, because it was necessary if the unions themselves were 'to survive the impatience and genuine determination by our new nations to move ahead'.[34] If they appeared to get in the way of 'the new governments', the unions would 'stand accused of either being foreign agents or imperialist agents, or of being just negative'.[35] The trade unions, in other words, occupied too strategic a position in the economy to be allowed to act with the aggression that nationalist leaders had usually applauded under colonial regimes. It followed that the new governments had to be closely concerned with the trade unions and, indeed, regulate their activities so far as they had a bearing on economic development.

Mboya declared his opposition to 'excessive intervention' by the state; and so a government ought to be condemned if it 'denies trade unions their existence' or (more weakly) if it 'tries to

[31] Perón, *Peronist Doctrine*, p. 352.

[32] Mboya, *Freedom and After*, p. 249; see also Onuoha, *The Elements of African Socialism*, pp. 132–3.

[33] See Goldsworthy, *Tom Mboya*.

[34] Mboya, *Freedom and After*, p. 250.

[35] Ibid. 192.

subjugate them to government policies without consultation or seeking cooperation'.[36] The minimal nature of these qualifications, and Mboya's broad interpretation of what development required, are indicated by his contention that an African government ought to provide 'some guidance on the question of association, because lack of guidance may lead to a disruption in the government's economic efforts'.[37] What he had particularly in mind here (as 'one example of the changes I believe are necessary') was government enforcement of the principle that there should be 'only . . . one central [trade union] organization . . . and that all trade unions should affiliate to it'.[38] As with Perón, unity and the exclusion of foreign influences were Mboya's justification.

> This [principle] means that any differences among trade unionists must be resolved within the movement itself, rather than by . . . organising opposition political parties or by siding with an opposition party which would simply be using the workers as a political weapon against the government. This guidance also eliminates the possibility of splinter groups or national centres of trade unions being formed at the instigation and with the money of foreign powers.[39]

There was, too, another justification of government intervention in trade union affairs: the need to protect trade union leaders against their own members. For the business of exercising restraint in line with governmental development policies, as another East African trade union leader confirmed, 'is often . . . unpopular and difficult'.[40] Governments, Mboya thought, were therefore under an obligation to 'help protect the very life of trade unions that risk their popularity by being responsible'.[41] But in the end, it seems, the fundamental justification of the state–trade union relationship that Mboya advocated amounted, in the case of a new African state intent on 'nation-building', to his belief that 'there is no essential conflict between unions and government'.[42]

ENGLISH FABIAN SOCIALISM

Only two of the contributors to the original *Fabian Essays*, published by the fledgling Fabian Society in 1889, so much as

36 Ibid. 248. 37 Ibid. 249. 38 Ibid.
39 Ibid. 40 Kamaliza, 'Tanganyika's View of Labour's Role', p. 10.
41 Mboya, *The Challenge of Nationhood*, p. 71.
42 Mboya, *Freedom and After*, p. 248.

mentioned trade unions. Graham Wallas and George Bernard Shaw, displaying a characteristically Fabian concern, both remarked that trade unions encouraged inefficient working practices. But Shaw perceived at least some virtue in them: he acknowledged that their 'value ... in awakening the social conscience of the skilled workers was immense'.[43] A third contributor, Sidney Webb, admitted three decades later that the essayists had paid insufficient attention to trade unionism.[44] He was, however, able to point out that the oversight had been corrected soon after by two path-breaking books which he and Beatrice Webb had written: *The History of Trade Unionism* (1894) and *Industrial Democracy* (1897).

It was in the final chapter of *Industrial Democracy* that the Webbs ventured to 'pass over into precept and prophecy' in order to set out their perception of the role that trade unions would play under the Fabian version of socialism.[45] Trade unionism, they thought, would constitute 'an essential organ of the democratic state' of the future, and would thus be 'a definitely recognised institution of public utility' with a membership that was 'virtually or actually compulsory'.[46] But it would also be subject to closer regulation. This meant, in particular, that the unions would be obliged to accept 'the fundamental condition that the business of the community must not be interfered with'—in other words, limitations on the right to strike and, ultimately, the settlement of serious industrial disputes by 'an authoritative fiat' of the state.[47]

A two-month visit to the Soviet Union, in the summer of 1932, convinced the Webbs (although they seem to have needed little convincing by that time)[48] that they had at last seen, in its essentials, the state that they could only envision in 1897. They recounted and elaborated what they saw in *Soviet Communism: A New Civilization?* One of the things they discovered, after touring Stalin's Russia, was that the 'soviet trade union ... is not formed to fight anybody'.[49] Instead of being an organ of conflict, as under capitalism, it was an organ of co-operation. At the same time, like the British trade union, it was also 'emphatically the organ of the

[43] Shaw *et al.*, *Fabian Essays*, p. 219.
[44] Ibid. ('Introduction to the 1920 Reprint'), p. 272.
[45] Webb, *Industrial Democracy*, p. 809.
[46] Ibid. 828. [47] Ibid. 813–4.
[48] See MacKenzie in Webb, *The Diary of Beatrice Webb*, iv. 270–2.
[49] Webb, *Soviet Communism*, p. 173.

wage-earners as such'.[50] And its extremely intimate links with party and state, of which the Webbs were well aware, did not persuade them to modify their opinion on that issue.

RUSSIAN COMMUNISM

Russian trade unions, the speculations of English Fabians apart, moved into virgin doctrinal territory when the Bolsheviks won power in October 1917. The works of Marx and Engels and the pre-revolutionary writings of Lenin provided no real clues to their role under socialism. The issue was to occasion a ferocious debate within the Communist party during the winter of 1920–1—and a politically effective resolution of it. But, until that time, confusion reigned.

Initially, the nature of the trade unions' relationship with the state was regarded as the heart of the matter. The 1st All-Russia Congress of Trade Unions, meeting in January 1918 with Bolsheviks in a two-to-one majority, affirmed that 'the trade unions will inevitably be transformed into organs of the socialist state', as far as the future was concerned; and, in the meantime, had to shoulder 'the chief burden of . . . rehabilitating the country's shattered productive resources'.[51] The subsequent outbreak of civil war, and the institution of 'war communism', also thrust the trade unions into the role of military recruitment and supply agencies. The 'statization' prediction of 1918 was endorsed, in January 1919, by the 2nd All-Russia Congress of Trade Unions, in which the Bolsheviks held a three-to-one majority.

Two months later, the Communist party took up a formal position on the issue for the first time. Its 8th congress declared, in distinctly syndicalist terms, that the unions 'must actually concentrate in their hands the entire administration of the whole public economy'.[52] The 9th congress, meeting in March 1920 (with the civil war ended and the Polish war still to come), corrected course by resolving that 'the trade unions must be gradually converted into auxiliary organs of the proletarian state, and not the other way around'.[53]

Leon Trotsky then became the polarizer. During 1920 he regenerated the railway system by the use of draconian methods

[50] Ibid. 218. [51] Quoted in Carr, *The Bolshevik Revolution*, p. 106.
[52] In McNeal, *Resolutions and Decisions* ii. 66. [53] Ibid. 101.

which included purging the entire leadership of the railway workers' union and other transport unions.[54] In November he evoked widespread antagonism by calling for a general trade union 'shake-up' along the lines of his treatment of the transport unions. By December, the party leaders themselves were so badly splintered on the issue that the central committee decided to leave it to the 10th party congress in March 1921; and, in the meantime, to allow open discussion on it. The fiery public debate that preoccupied the meetings and publications of the Communist party during the following three months was the first and the last of its kind that the party experienced.[55]

The dispute focused on the question of whether the trade unions should regard their primary task as the promotion of production or the protection of their members' interests. It was assumed that the productionist role entailed subordination to the state, whereas the protectionist role did not. The contending positions were eventually reduced to three. The productionists, led by Trotsky and Nikolai Bukharin, advocated the complete 'statization' of the unions (their 'transformation . . . into production unions'),[56] arguing that the socialist state itself would automatically protect workers' interests. The protectionist 'workers' opposition', led by A. G. Shlyapnikov and Alexandra Kollontai, advanced a basically syndicalist solution. The middle ground was held by 'the ten', who included Lenin, Josef Stalin, and Mikhail Tomsky, leader of the central trade union body. Their draft resolution proposed, in essence, that the trade unions should be concerned with *both* production and protection. It was overwhelmingly endorsed by the congress.

The 10th congress was significant for other, and weightier, reasons as well.[57] Nevertheless, the debate which preceded it revolved almost entirely around the trade union issue. Lenin's contributions to it, all made in little more that a month during December 1920 and January 1921, consisted of three speeches and a pamphlet (*Once Again on the Trade Unions*). These are the source of his commonly cited remarks on trade unions under socialism.

[54] See Deutscher, *The Prophet Armed*, pp. 498–502.

[55] For a more precise indication of the lively and unusual nature of this debate, see Carr, *The Bolshevik Revolution*, p. 223 n.

[56] Trotsky, quoted in Carr, ibid. 225.

[57] See McNeal, *Resolutions and Decisions*, ii. 124.

His basic assumption was that trade unions had a vital role to play during the 'dictatorship of the proletariat', the socialist phase. Though created by capitalism, they were inseparable from socialism. Whether they had a future beyond socialism was another matter: 'it will be up to our grandchildren to discuss that'.[58] In the meantime, however, they had 'an extremely important part to play at every step of the dictatorship of the proletariat'.[59] So important, indeed, that the 'dictatorship cannot be exercised . . . without a foundation such as the trade unions'.[60] And, even more imposing: 'the trade unions are . . . the source of all our power'.[61] The conclusion was plain. It was imperative that the party got its relationship with the unions right. If they fell out with each other, it spelt 'certain doom for the Soviet power' and would 'topple the dictatorship of the proletariat'.[62]

WHOSE INTERESTS?

For the authoritarian, the primary responsibility of trade unions is to the state. Other entities may be mentioned (such as the nation, the community, the common good, the national interest, the working masses) but, in the end, they all boil down to the state. Similarly, the authoritarian often specifies an ostensibly qualifying responsibility relating to union members or discrete groups of workers. Invariably, however, there is the ultimate assertion of a final and overriding identity between the 'genuine' interests of workers and the interests represented by the state, an assertion which denies a trade union the autonomy necessary to give substance to the qualification.

Mussolini's Law of 3 April 1926 (No. 536) assured a trade union that it would obtain legal recognition provided that, among other things, its leading official could guarantee not only his 'ability' and 'morality', but also his 'sound national loyalty'.[63] The loyalty thus required of trade union leaders (and, by necessary implication, their members and organizations) involved a responsibility direct to the state, given the Fascist perception of the nation as 'an organism . . . which finds its integral realization in the Fascist State'.[64]

[58] Lenin, *Collected Works*, xxxii. 23.
[59] Ibid. 20. [60] Ibid. [61] Ibid. 58.
[62] Ibid. 58, 83. [63] Mussolini, *Fascism*, p. 76. [64] Ibid. 133.

Hitler, when he took over and reshaped the German trade union movement to his satisfaction, postulated an identical responsibility. The 'National Socialist State' was the focus for the Labour Front.[65]

Perón was less explicit on the issue of responsibility. But the implication was the same, and clear enough, when he spoke of the necessity for 'united and well-directed labour unions, because inorganic masses are always the most dangerous to the State and to themselves'.[66]

Mboya, for his part, reached back to African traditional life for the model the trade unions should follow. He extolled the 'sense of family and the sense of community' that were part of the 'African heritage': 'This spirit of mutual social responsibility must be adopted by our trade unions.'[67] And of course, as we have already seen, he perceived 'no essential conflict between unions and government'.[68] Jomo Kenyatta, president of independent Kenya, made the point in another way: 'The first responsibility of the unions must be to develop a disciplined . . . and responsible labour force.'[69]

Sidney and Beatrice Webb initially took a different line by espousing a carefully qualified notion of trade union responsibility under socialism. Even the most democratic state, they thought, would not be flawless, and so trade unions would still have a responsibility to their members alone—to protect them *against* the state on the occasions when protection was required.[70] That was their belief in the 1890s. In the very different world of the 1930s, however, the Webbs beamed on trade unions which, without qualification, were obliged 'to protect . . . not so much the wage-rates of the workers in particular industries, as the . . . whole conditions of life . . . of all the wage earners of the USSR'.[71] That obligation, given the Communist party's penetration of both the state and the trade unions (which the Webbs clearly understood),[72] amounted to a responsibility to the state. Indeed, they specifically identified Soviet trade unionism as a state agency when they

[65] Cited in Brady, *The Spirit and Structure of German Fascism*, pp. 120–1.
[66] Perón, *Peronist Doctrine*, p. 348.
[67] Mboya, *The Challenge of Nationhood*, p. 71.
[68] Mboya, *Freedom and After*, p. 248.
[69] Quoted in Beling, *The Role of Labor in African Nation Building*, p. 23.
[70] See Webb, *Industrial Democracy*, p. 825.
[71] Webb, *Soviet Communism*, p. 172.
[72] See ibid. 350–62.

described it as 'closely connected with *the other* organs of the state'.[73]

There is a strong whiff of equivocation about Lenin's public statements with a bearing on trade union responsibility under socialism. During the winter of 1920–1, in the debate preceding the 10th party congress, he rejected Trotsky's exclusive emphasis on the interests of the state. In particular, he dismissed Trotsky's supporting argument that there could be no conflict between the workers' interests and those of the 'workers' state'. Lenin described this as 'an ideal we shall achieve in 15 or 20 years time', at best.[74] For the state was still 'not quite a workers' state'; but rather 'a workers' state *with a bureaucratic twist to it*'.[75] That fact imposed on the trade unions a double responsibility. They had, at once, 'to protect the workers from their state, and to get them to protect our state'.[76] Lenin reaffirmed the first, and narrower, responsibility of the trade unions in 1922 when the New Economic Policy was expanded to admit an element of free enterprise into the Soviet economy and to subject state industry to market pressures. Trade unions operating in private enterprises, obviously enough, were charged with protecting 'in every way the class interests' of workers.[77] Less predictably, unions involved with *state* enterprises had a similar, if not identically stated, 'duty . . . to protect the interests of the working people'.[78]

Superficially, this may seem to amount to much the same kind of exclusively class responsibility as that which Lenin, along with Marx and Engels, had ascribed to trade unions under capitalism. But, as Lenin took particular pains to point out in 1922, the interests of the working class were vastly different under socialism than they had been under capitalism. 'Following its seizure of political power, the principal and fundamental interest of the proletariat lies in securing an enormous increase in the productive forces of society and in the output of manufactured goods.'[79] And what this meant (although, in the circumstances of the time, it seems that Lenin could not afford to be too blunt about it) was that trade unions, as Trotsky had maintained, had an ultimately overriding responsibility to the state.

[73] Ibid. 218. Emphasis added.
[74] Lenin, *Collected Works*, xxxii. 24.
[75] Ibid.
[76] Ibid. 25.
[77] Ibid. xxxiii. 185.
[78] Ibid. 186.
[79] Ibid. 188–9.

Hence . . . the task of the trade unions is to facilitate the speediest and smoothest settlement of these disputes to the maximum advantage of the groups of workers they represent, *taking care, however, not to prejudice the interests of other groups of workers and the development of the workers' state and its economy as a whole*.[80]

Much later, in Stalinist times, this broad responsibility was stated more flatly by Alexander Lozovsky. His topic was the role of Communist party cadres in Soviet trade unions. Their task was to persuade non-party unionists that proposals emanating from the party were 'correct', and correctness was determined by what was good for 'both the given collective and *society as a whole*'.[81]

INTERESTS IN WHAT?

Authoritarian theories of trade union purpose invariably end up with production as the overriding objective. One symptom of that is the great weight they place on industrial peace and labour discipline. Typically, too, they stress the capacity of the state to act as the sole protector of workers' legitimate industrial interests.

Mussolini

The essential purpose Mussolini attributed to Fascist unions was made crystal clear by Law No. 536 of 1926. This specified that a trade union, in order to obtain legal recognition, had to 'prove' compliance with the requirement that, in addition to protecting 'the economic and moral interests of its members', it promoted their 'moral and patriotic education'.[82] And the purpose that morality and patriotism implied was higher production. Thus the Fascist 'Charter of Labour' referred to the 'subordination' of the sectional interests of employers and employees 'to the superior interests of production'.[83] The function this involved for trade unions, as Law No. 536 made abundantly clear, was an educational function.

Hitler

A 'National Socialist Trades Union', Hitler (pre-1933) acknowledged, would be concerned with the 'defence and representation of

[80] Lenin, *Collected Works*, xxxiii. 187. Emphasis added.
[81] Lozovsky, *Handbook on the Soviet Trade Unions*, p. 121. Emphasis added.
[82] Mussolini, *Fascism*, p. 76.
[83] In Oakeshott, *Social and Political Doctrines of Contemporary Europe*, p. 184.

the workers', short of entanglement in the 'class war'.[84] But it also had to share the awareness expected of the properly motivated worker, 'that the nation's prosperity means material happiness to himself'.[85] In other words, Hitler envisaged trade unions preoccupied with increasing production. They would also be intimately linked to the state, a point later illustrated with particular force in the German Law of 20 January 1934 which specified that a worker, to be eligible for appointment as a 'Spokesman' (resembling, in a formal sense, a shop steward), had to be 'wholly dependable with regard to his willingness at all times unreservedly to uphold the national State'.[86] This echoed Hitler's conception of the central purpose of the German Labour Front itself. It was, he declared, less concerned with 'the material questions of daily labour life' than with the 'high aim' of 'educating all Germans active in labour life to the National Socialist State and to the National Socialist Doctrines'.[87] The central importance he attached to an educational function could not be made more obvious.

Perón

The protection of workers' interests, as a trade union purpose, was often on Perón's lips. At the same time, however, 'responsible' trade unions, 'logical and rational organisms, well directed', would also prevent 'the masses . . . from going too far in their demands'.[88] 'Real protection for the worker', in any case, depended upon a trade unionism which 'works together with . . . the State'.[89] And what working with the state came down to, in the end, was 'the success of production', to which workers had 'the responsibility of contributing by their labour and *the support of their organizations*'.[90] Thus, Perón pushed the slogan, 'Produce! Produce! Produce!' and spoke tirelessly of the need 'to urge the working people to produce'.[91] He advocated 'labour discipline' ('without it fruitful labour is impossible') and he deplored industrial disputes, which were 'always destructive'.[92] The trade unions had a critical role to play in all this. It was up to them, as part of their 'daily mission', to

[84] Hitler, *My Struggle*, p. 238. [85] Ibid. 239.
[86] In Oakeshott, *Social and Political Doctrines of Contemporary Europe*, p. 214.
[87] Cited in Brady, *The Spirit and Structure of German Fascism*, pp. 120–1.
[88] Perón, *Peronist Doctrine*, p. 309.
[89] Perón, *Peron Expounds His Doctrine*, p. 261.
[90] Perón, *Peronist Doctrine*, p. 325. Emphasis added.
[91] Ibid. 227, 238. [92] Ibid. 308, 312.

'convince each of the men working in this country, of the need to work, construct, and produce'.[93] In other words, they were allotted the major purpose of promoting production and, consequently, a function that was above all educational.

Mboya

The primacy of production was equally clear to Mboya. He accepted the protection of workers' interests as a proper trade union purpose, but denied that it could, in the African context, be the only purpose. 'It is possible for trade unions to fulfil two purposes in Africa: they can defend the rights and promote the interests of the workers, and at the same time cooperate with the government in economic reconstruction.'[94] The burden of his argument, however, was that the first purpose depended on the second, which therefore had to be given decisive priority in a developing country. And 'economic reconstruction' required precisely 'sacrifices, or self-discipline' which necessarily impinged on the interests of trade unions and their members.[95] So the strike weapon had to go. Strikes and lockouts 'cannot be permitted in poor developing countries . . . the social costs are unbearable'.[96] The trade unions, instead, should turn their attention to 'developing a work force that acts responsibly on policy matters and produces profitably at the workbench'.[97] Again, there is in this the attribution of an educational function. It included 'training . . . members both in the skills of their work and in the importance of their contribution to their country's development . . . and . . . improving the discipline of workers'.[98] More broadly, trade unionism could also be 'a very useful instrument of education against negative tribalism'[99]—and, in general, as Mboya's trade union successor pointed out, could train 'our people to become better citizens'.[100]

The Webbs

From the perspective of the 1890s, Sidney and Beatrice Webb predicted that the role of trade unions in the socialist state of the

[93] Perón, *Peronist Doctrine*, p. 327. [94] Mboya, *Freedom and After*, p. 248.
[95] Ibid. 249. [96] Mboya, *The Challenge of Nationhood*, pp. 70–1.
[97] Ibid. 71. [98] Ibid. 178.
[99] Mboya, *Freedom and After*, p. 71.
[100] Lubembe, 'Trade Unions and Nation Building', p. 21.

future would depend, above all, on a protective purpose, stemming from their singular capacity for expressing the needs and grievances of workers. Certainly, they would have other purposes as well; and they would have to accept that in most things social efficiency, rather than the gratification of sectional interests, was the principal consideration. Nevertheless, when the Webbs wanted to explain why 'Trade Unionism is not merely an incident of the present phase of capitalist industry, but has a permanent function to fulfil in the democratic state',[101] it was the protective purpose on which they leant most heavily.

For even under the most complete Collectivism, the directors of ... industry would ... remain biassed in favour of cheapening production, and could, as brainworkers, never be personally conscious of the conditions of the manual laborers. And ... experience of all administration on a large scale ... indicates how difficult it must always be, in any complicated organisation, for an isolated individual sufferer to obtain redress against the malice, caprice, or simple heedlessness of his official superior. Even a whole ... grade of workers would find it practically impossible, without forming some ... association of its own, to bring its special needs to the notice of public opinion ... In short, it is essential that each ... section of producers should be at least so well organised that it can compel public opinion to listen to its claims, and so strongly combined that it could if need be, as a last resort against bureaucratic stupidity or official oppression, enforce its demands by a concerted abstention from work, against every authority short of a decision of the public tribunals, or a deliberate judgement of the Representative Assembly itself.[102]

Thus, even under socialism, a measure of genuine trade union autonomy was imperative to provide workers with a shield against certain 'bureaucratic stupidity or official oppression'.

In the 1930s, confronted with an actual socialist state, the Webbs described the purpose of the Soviet trade union in terms which nominally retained the protective purpose (consistent with their famous definition of a trade union), but also laid down a striking qualification:

its essential [purpose] is that of maintaining and improving the worker's conditions of life—*taking, however, the broadest view of these, and seeking their advancement only in common with those of the whole community of workers.*[103]

[101] Webb, *Industrial Democracy*, p. 823.
[102] Ibid. 824–5.
[103] Webb, *Soviet Communism*, p. 218. Emphasis added.

The qualification implies that the conditions of one group of workers should not be improved without improving the conditions of all workers. The overriding purpose of promoting production flows axiomatically from this. There is no doubt that it did so for the Webbs. And the acceptance of such a purpose, as they saw it, made for an unusual, but unquestionably commendable, trade union movement.

> It is, we think, unique in the intense interest that it takes in increasing the productivity of the nation's industry . . . and in . . . how cordially it has accepted the various arrangements . . . for securing the utmost possible output at the lowest possible expense to the community.[104]

In a 'postscript' to the second edition of *Soviet Communism* (in which the question mark was dropped from the sub-title: *A New Civilization*), the Webbs wrote glowingly of 'an emotional passion for production' which they claimed to have detected among Soviet workers—a passion demonstrated, so they judged, by the fact that 'it is the trade unions that most strongly insist on the utmost use of labour-saving machinery and piece-work rates, socialist competition and Stakhanovism'.[105]

The Webb's applause for the great weight which Soviet trade unions undoubtedly placed on the production purpose was not, in itself, at all out of keeping with the views that they had expressed earlier in *Industrial Democracy*. The disjunction arose because of what was *omitted* from their commentary on Soviet trade unionism and its purposes. For the Webbs made no reference to the protective purpose which they had once thought essential, 'even under the most complete Collectivism', because of inevitable differences between managers and managed in large organizations. They had come, it appeared, to the narrower conclusion that social efficiency required trade unions concerned exclusively with the purpose of promoting production and so an educational function.

Lenin and Stalin

In the winter of 1920–1, Lenin was not similarly inclined to ignore the possibilities of the 'bureaucratic stupidity or official oppression' that the Webbs had foreseen in 1897.[106] Indeed, this was precisely

[104] Webb, *Soviet Communism*, p. 218. [105] Ibid., 2nd edn. (1937), p. 1215.
[106] Webb, *Industrial Democracy*, p. 825.

the ground on which he justified the notion of a dual trade union purpose under socialism—and, in particular, the proposition that unions retained the protective purpose they had under capitalism. The point had, in fact, been formally made in one way as early as 1919, in a resolution of the 8th party congress which described the trade unions as 'the principal means of struggle against bureaucratization in the economic apparatus of the Soviet government'.[107] Lenin elaborated it in the winter debate. He rejected, as 'a hasty exaggeration', Trotsky's argument that trade unionism's protective purpose had disappeared along with the struggle against the capitalist class.[108] It was true, he said, that the trade unions were no longer involved in 'the *class* economic struggle', but they now faced what he described as 'the *non-class* "economic struggle" '; and this committed them to 'combating bureaucratic distortions' in the state, and to 'safeguarding the working people's material and spiritual interests' in ways that were beyond the state.[109] In short, the trade unions had a protective purpose which required them 'to stand up for the material and spiritual interests of the working class' against a state that was, for the time being, 'a workers' state with bureaucratic distortions'.[110] And this situation, this 'struggle' on the part of the unions, would continue 'for many more years to come'.[111] It followed that, for years to come, the unions would need a measure of independence from the state in order to fulfil their protective purpose.

There can be no question, however, that it was the other purpose of promoting production which was of greater importance in Lenin's eyes. This was evident in the way he handled the two purposes, as a matter of location, elaboration, and emphasis in his contributions to the winter debate. But, of course, he diverged from Trotsky, in this debate, by denying that promoting production was the *only* purpose of a trade union—a denial that extended also to the unions' use of coercion as a means to that end. Both Lenin and Trotsky, that is to say, took Trotsky's 'statization' proposal to imply not only that the trade unions would focus exclusively on production (as against protection), but that they would employ the coercive powers of the state in pursuit of increased production. Lenin's position was that coercion was inappropriate as a trade

[107] In McNeal, *Resolutions and Decisions*, ii. 66.
[108] Lenin, *Collected Works*, xxxii. 100.
[109] Ibid. [110] Ibid. 24, 48. [111] Ibid. 100.

union device under socialism, on the ground that the peculiar value of the trade unions stemmed precisely from the effectiveness of the non-coercive means available to them.

'Persuasion', Lenin told the 10th party congress, 'must come before coercion. We must make every effort to persuade people before applying coercion.'[112] And persuasion, as he had been at pains to make clear throughout the winter debate, required the existence of 'a number of "transmission belts" running from the vanguard [the party] to the mass of the advanced class, and from the latter to the mass of the working people'.[113] The trade unions were such transmission belts. They were 'a *link* between the vanguard and the masses', and critical components of an 'intricate transmission system'.[114] They were the answer to the key question of 'how to approach the mass, win it over, and keep in touch with it'.[115] And although they were pre-eminently organizations of the new ruling class which was 'exercising coercion through the state', under the dictatorship of the proletariat, they were not themselves 'a state organization'.[116] Their function was quite different. For they were not 'designed for coercion, but for education'.[117]

The educational function of trade unions under socialism had two aspects. On the one hand, it involved ideological training. As Lenin put it, the trade union became a 'school of communism'[118] (a term applied to trade unions earlier, in 1920, by way of a resolution of the 9th party congress),[119] a 'school of unity, solidarity'.[120] On the other hand, the educational function involved elements of technical training; and Lenin described unions also as 'a school of technical and administrative management of production'.[121] This, however, seems not to have meant formal training. The unions were 'a very unusual type of school, because there are no teachers and no pupils'; and it was 'by their daily work' that they 'bring conviction to the masses'.[122]

The production purpose was plainly central to Lenin's depiction of trade unions as 'a school of communism and a school of management'[123]—or, a little more elaborately, 'a school of administration, a school of economic management, a school of communism'.[124] On one occasion, however, he expressly associated the

[112] Lenin, *Collected Works*, xxxii. 212. [113] Ibid. 21. [114] Ibid. 20, 23.
[115] Ibid. 23. [116] Ibid. 20. [117] Ibid.
[118] Ibid. [119] See McNeal, *Resolutions and Decisions*, ii. 101.
[120] Lenin, *Collected Works*, xxxii. 96. [121] Ibid.
[122] Ibid. 20. [123] Ibid. 98. [124] Ibid. 20.

protective purpose with the educational function by describing unions as 'a school' in which, as well, 'you learn how to protect your interests'.[125]

The 10th party congress, on 16 March 1921, adopted a resolution on the trade unions which echoed Lenin's views. It specified the educational function of the unions as '*schools of communism* . . . a primary school to provide organizational skills and political training for the broadest . . . masses of the working people'.[126] And it placed the emphasis firmly on the production purpose: the test of 'a true school of communism' was the ability 'to enlist the most backward strata of the working people in the work of consciously improving Soviet economy'.[127] There was, nevertheless, a brief reference to the protective purpose in that unions were urged to be sensitive in their 'practical work', even 'at times to the most minor . . . questions of the everyday life of the working masses'.[128]

The conception of a dual trade union purpose, which Lenin expounded in the winter of 1920–1, has commonly been treated as if it were all that he had to say on the issue of union purpose under socialism. It is a conception which (with its admission of the protection of workers' interests against 'bureaucratic distortions' as a legitimate goal) carries the apparent implication, as we have seen, of trade unions which have the capacity to act independently of the socialist state. Interpreted in this way, Lenin's dual purpose theory is not an authoritarian doctrine, as that term was defined at the outset of this chapter. But such an interpretation of Lenin's position is too simple because, on the one hand, it neglects what he had to say on the subject both before and after the winter debate; and on the other, it overlooks a crucial assumption about the role of the party which underlay his contributions to that debate.

In January 1919 Lenin had spoken in support of the resolution (mentioned above) in which the 2nd All-Russia Congress of Trade Unions affirmed the future 'statization' of the unions. He condemned the 'notorious independence slogan', peddled by Mensheviks and others, but also rejected the possibility of merging the trade unions and the state 'at once, at one stroke'.[129] In the meantime, the trade unions' main task was to 'teach more and more people the art of state administration'[130]—that is, encourage

[125] Ibid. 96. [126] In McNeal, *Resolutions and Decisions*, ii. 126.
[127] Ibid. 127. [128] Ibid. 129.
[129] Lenin, *Collected Works*, xxviii. 412, 425. [130] Ibid. 427.

production. A year later, in the desperate period that followed the end of the civil war, he was speaking (though not to a specifically trade union audience) of the need to create a 'labour army'.[131] In April 1920, the 9th party congress affirmed that 'the trade unions must be gradually converted into auxiliary organs of the proletarian state'; and, after mentioning their 'educational' function, described as 'their most important function, the economic-administrative function'.[132] Lenin did not speak to this resolution, but there was no question of his support for it. About the same time, he urged an audience of mineworkers' delegates 'to create a firm labour discipline, raise the productivity of labour and foster the spirit of self-sacrifice among the workers in the coal industry'.[133] His theme was the same a day or two afterwards, when he spoke at the 3rd All-Russia Congress of Trade Unions. The 'economic development of socialism' depended above all on 'labour discipline', and on a readiness to 'work with military determination . . . brushing aside all group and craft interests, sacrificing all private interests'.[134]

There was nothing at all in these utterances about a dual purpose, or about trade unions protecting workers' interests. On the contrary, the emphasis was completely on the abnegation of workers' special interests, and their subordination to the state's interest in production. Lenin, in other words, seems to have had as one-sided a view of trade union purpose as Trotsky, until Trotsky's reckless threat to 'shake up' the unions set a match to the issue. Only then, in the winter of 1920–1, did Lenin talk of a dual trade union purpose under socialism.

He produced one other major statement on the issue. Early in January 1922 he finished writing *The Role and Functions of the Trade Unions under the New Economic Policy*. This complex statement, adopted by the Communist party's central committee and its 11th congress, was prompted by the new situation arising from the extension of private enterprise into the industrial sector. It opened with a more elaborate account of the *protective* purpose of trade unions than he had found it necessary to provide a year earlier. First, in private enterprises, it was essentially business as before the revolution: the trade unions had to protect 'the

131 Lenin, *Collected Works*, xxx. 312.

132 In McNeal, *Resolutions and Decisions*, ii. 100–2.

133 Lenin, *Collected Works*, xxx. 500–1.

134 Ibid. 504, 514.

proletariat in its struggle against capital'.[135] Second, in state enterprises, which were being moved on to a profit-and-loss basis under the NEP, they had similarly to protect workers against 'the blunders and excesses of business organisations resulting from bureaucratic distortions of the state apparatus'.[136] In both cases, the protective purpose even justified a right to strike, so long as it was used only to curb 'the class appetites of the capitalists', in the first case, and to combat 'bureaucratic distortions' in the second.[137] Nevertheless, in relation to state enterprises at least, the right to strike was subject to considerable additional qualification, the flavour of which is conveyed in the remark that an 'infallible' test of the success of trade unions was the extent to which 'they succeed in averting mass disputes in state enterprises'.[138]

Lenin's statement then moved on to the *production* purpose. This was not to involve (and the point was made most forcefully) anything like 'direct interference' by trade unions in factory management.[139] What it entailed was an educational function. As a 'school of communism in general', the trade unions had the task of training workers 'in the art of managing socialist industry'.[140] He listed a number of specific ways in which, with this end in mind, the trade unions 'must collaborate closely and constantly with government'—including, among others, taking part in state planning bodies; selecting and training workers for managerial responsibilities; running 'disciplinary courts' to enforce labour discipline; and participating in 'production propaganda'.[141]

The most remarkable feature of Lenin's 1922 statement, however, was the way in which, for the first time, he canvassed the problem at the heart of the dual purpose conception—the inevitable tension between the purposes of protection and production. There were, as he put it, 'a number of contradictions in the various functions of the trade unions':

On the one hand, their principal method of operation is that of persuasion and education; on the other hand, as participants in the exercise of state power *they cannot refuse to share in coercion*. On the one hand, their main function is to protect the interests of the masses of the working people in the most direct and immediate sense of the term; on the other hand, as participants in the exercise of state power . . . *they cannot refuse to resort*

[135] Ibid. xxxiii. 185.
[136] Ibid. 186.
[137] Ibid. 187.
[138] Ibid. 188.
[139] Ibid. 189.
[140] Ibid. 190.
[141] Ibid. 190–1.

to pressure. On the one hand, *they must operate in military fashion* . . . on the other hand, specifically military methods of operation are least of all applicable to the trade unions . . . These contradictions are no accident, and they will persist for several decades.[142]

The italicized phrases in this quotation leave little doubt about the way in which Lenin considered such 'contradictions' should be resolved as a matter of practical operation. They stripped from the dual purpose conception the ostensible even-handedness with which it was clothed when he introduced it in the winter of 1920–1. Yet even then, during the winter debate, the way in which 'contradictions' would be resolved was, to all intents and purposes, a foregone conclusion. For a while he had implied a measure of trade union independence of the *state*, in attributing a protective purpose to unions, there was no question of their being independent of the *party*. The significance of this was that any resolution of 'contradictions' would be determined by the party, not the unions as such. Party control of the trade unions, that is to say, was the crucial point; and conceptions of trade union purpose were secondary. Leninist doctrine never deviated an inch on the issue of party control.

The metaphor of the 'transmission belt', which Lenin coined to describe the unions' relationship with the party,[143] caught his meaning precisely. As he explained in mid-1920, although the trade unions were 'formally *non-Party*', in reality 'the directing bodies' of almost all union organizations were 'made up of Communists and carry out the directives of the Party'.[144] The tail of the resolution in which the 10th party congress enshrined his conception of a dual trade union purpose, in March 1921, carried the same message: 'The Russian Communist Party . . . unquestionably directs . . . the entire ideological side of the work of the trade unions.'[145] The point, in any case, was already made, in brutal detail, by the party's own formal rules.

The Central Committee directs the work of the central . . . social organisations [including the trade union centre] through the party fractions . . . The guberniia [provincial] committee directs the activities of . . . the trade unions . . . through the corresponding party fractions . . . Fractions, regardless of their importance are entirely subordinate to the party.[146]

[142] Lenin, *Collected Works*, xxxiii. 193. Emphasis added.
[143] Ibid. xxx. 478.
[144] Ibid. xxxi. 48.
[145] In McNeal, *Resolutions and Decisions*, ii. 128.
[146] Ibid. 93–4, 97.

And in his 1922 statement, Lenin expressly applied the point to the 'contradictions' which were inherent in the dual purpose conception: 'the afore-mentioned contradictions will inevitably give rise to disputes, disagreements, friction, etc. A higher body is required with sufficient authority to settle these at once. This higher body is the Communist party.'[147]

Thus the 16th party congress of 1930, the first to register Stalin's ascendancy, was asserting a Leninist principle—not endorsing a Stalinist departure—when it applauded Stalin's dismissal of Tomsky and other 'anti-Leninist' trade union leaders on the ground that they were guilty of 'weakening party leadership of the trade union movement and . . . opposing the trade unions to the party'.[148] Similarly, Kaganovich's later litany of the purged union leaders' errors was well within the parameters of Lenin's dual purpose conception:

> Opposing . . . the workers' interests to the interests of Socialist industry, and the unions' defensive functions to their production work, they either kept quiet about the unions' role as a school of Communism, or interpreted it . . . one-sidedly as a school of association, a school of defence of one's interests.[149]

Stalin, despite the claims of sycophants,[150] added nothing of importance to the theory of trade unions under socialism. He weighed into the winter debate of 1920–1 with a *Pravda* article attacking Trotsky's advocacy (as Stalin put it) of 'the method of *coercion* (the military method)', and arguing in favour of 'the method of *persuasion* (the trade-union method)'.[151] There is a terrible irony (given the policies he was to apply under the five-year plans) in his assertion that it was 'impossible' for the 'firm confidence' of the working masses to be won 'by coercive methods and by "shaking up" [Trotsky's term] the unions from above, for such methods . . . engender distrust of the Soviet power'.[152] Later, in *On the Problems of Leninism,* he merely restated the bones of Lenin's approach (with much quotation), if without once mentioning the protection of workers' interests as a union

[147] Lenin, *Collected Works*, xxx. 193.
[148] In McNeal, *Resolutions and Decisions*, iii. 80.
[149] Quoted in Conquest, *Industrial Workers in the USSR*, p. 152.
[150] See e.g. Lozovsky, *Marx and the Trade Unions*, p. 180.
[151] Stalin, *On the Opposition*, p. 2.
[152] Ibid. 10.

purpose.[153] But his theoretical position remained, throughout his long reign, as consistently Leninist as that of his successors. In neither case was it a difficult position to maintain in view of the latitude that Lenin had built into his theory of trade union purpose under socialism.

153 See Stalin, *Problems of Leninism*, pp. 165–8.

7

OF DEFINITIONS AND OUTSIDERS

Table 1 summarizes the leading features of the five categories into which conceptions of trade union purpose fall. There are some notable overlaps. One is the nomination of the working class as the interest to be served by both the syndicalists and the Marxist-Leninists. Another is the specification by organicists, as well as by pluralists and syndicalists, of independent involvement in industrial administration as a goal. Most striking, however, is the near-universal endorsement of some kind of educational function—with only the pluralists (and not all of them, as Perlman demonstrates) withholding an express blessing.

In the end, however, it is the disparities which predominate; and their implication for the Webbs' definition of a trade union is inescapable. They mean that the definition does not work in the way that the Webbs assumed. For of the two criteria they relied on, only that referring to the nature of a trade union's membership measures up to the requirements of a general empirical definition in the sense that it clearly delineates the trade union as a distinctive type of association.

Their second criterion, that of purpose, sets quite different boundaries. It is not (at least in the case of trade unions) an ideologically neutral criterion. On the contrary, unlike the nature of membership, it is precisely an expression of ideological preferences. The boundaries it sets, as a consequence, are not only different from those set by the nature of membership criterion; they are also narrower. Thus the Webbs' definition (relying on the purpose of 'maintaining or improving the conditions of their [members'] working lives') is consistent, on any reasonable interpretation, only with the pluralist position—and with trade unions espousing pluralist goals, functions, and responsibilities. By the same token, it is a definition which in application excludes much more than just associations which either make no claim to be trade unions or, if they do, fail to comply with the nature of membership criterion. For it excludes, as well, a host of associations which not only call

TABLE 1. The Theories of Trade Union Purpose

	Dimension 1 (Whose Interests?)	*Dimension 2* (Interests in What?)	
	Responsibility	*Goals* (layers)	*Functions*
Pluralist	Union members	*Job Regulation*; security and freedom on the job; participation in administration of industry	*Collective bargaining* (rule-making)
Syndicalist	Working class[a]	*Raise workers' consciousness*; overthrow capitalism; administer industry after revolution	*Education* (revolutionary and administrative); strike organization; administration
Marxist-Leninist	Working class	*Raise workers' consciousness*; help overthrow capitalism	*Education* (revolutionary); strike organization (party-directed)
Organicist	Society/Nation	*Promoting members' moral well-being*; promoting their material well-being; workers' participation, co-operative enterprises, profit-sharing, joint ownership	*Education* (moral improvement); collective bargaining
Authoritarian	State/Party	*Promoting production*; protecting workers' interests (chiefly or wholly by raising their productivity)	*Education* (for productivity); administrative participation (under state/party direction)

[a]It should be remembered that the Guild Socialists had, in effect, a more expansive conception of the class than the other syndicalists: they included in it all non-manual employees.

themselves trade unions (or something similar) but also have memberships that are confined to employees with some commonality of employment. And it excludes them on the basis that they do not profess as their principal purpose the one purpose which, it is implicitly claimed, a trade union *ought* to pursue. In other words, such a definition is ideologically selective, and therefore incapable of delineating an organizational type which observably exists and operates, with a variety of purposes, in the modern world.

This is not to suggest that definitions relying on purpose are intrinsically illegitimate. It depends on the use made of them. Such definitions have their place. George Bain and Robert Price illustrated this when they chose to define the trade unions covered by their study of union growth in eight countries with reference to a 'basic purpose' of representing the 'job interests' of members.[1] They were concerned with a particular kind of trade union and a particular kind of trade union movement, and their definition both acknowledged this and made it clear.

Nor is there any question that differences in trade union purpose involve differences of real significance. In accepting this, one can even go as far as Bain and Price who described trade unions which did not fit their definition as 'fundamentally different in nature'[2]—so long as this is not taken to mean that such organizations lack the essential quality of *genuine* trade unions, and are really usurping the title. The notion that the title itself, properly interpreted, implies a specific, preferred purpose is often evident in writings on trade unions.[3] My point is that this notion is unjustified, however 'fundamental' the ideological differences dividing trade unions. In other words, trade unionism *as an institutional form* is to be defined by the nature of its membership, not by its ascribed purpose.

OUTSIDERS AND TRADE UNION PURPOSE

One other feature to emerge from the preceding five chapters concerns the source of ideas about the goals and responsibilities of trade unions. In the general literature of organization theory, purpose has usually been treated as an empirical problem. The emphasis, in other words, has tended to be on the goals which an

[1] Bain and Price, *Profiles of Union Growth*, p. 2. [2] Ibid.
[3] See e.g. the references to Hyman and Fryer, 'Trade Unions', in ch. 1 above.

organization actually pursues and/or claims to pursue. What that comes down to is the purposes attributed to an organization by those who comprise it. Similarly, conflicts about organizational goals have usually been perceived, in the literature, in terms of divisions among the constituents of the organization concerned.

In the case of trade unions, however, their leaders and members have never exhausted the list of those with firm views about union purpose. Outsiders have been perennially eager to proclaim, and often to impose, opinions about the purposes that trade unions *ought* to pursue. Nor is this kind of concern with 'the definition of union goals', as Hyman and Fryer put it, confined to the 'employers, managers and politicians' who feel that 'their fortunes [are] affected by trade unionism and the actions of trade unionists'.[4]

From its beginnings, trade unionism in most countries has been blessed (or cursed, depending on the viewpoint) with non-workers expressing concern for its purposes. Middle-class intellectuals with an interest in social reform have been particularly prominent in this respect. The middle-class creators of the younger trade union movements that first emerged in the colonial territories of Africa and Asia are well known, and their doings often examined. But these movements differed radically from their older counterparts in this respect only in so far as the outsiders tended to become, and often remained, formal officials of the trade union organizations with which they were associated. Britain, for example, had its Robert Owen, the founder of the ill-fated Grand National Consolidated Trades Union of the 1830s, and himself an industrial employer. Later, there were the Positivists, like Frederic Harrison and Thomas Hughes, who gave guidance to the fledgling Trades Union Congress. And, of course, there was also Marx with his more internationally inclined trade union concerns. Elsewhere, in the Americas and in Continental Europe, east and west, there was similarly no lack of self-appointed middle-class advisers to emerging trade union movements. Moreover, it needs to be emphasized that not all of them came from the left of the political spectrum—though all had in common a professedly profound concern for the welfare and the 'true' interests of the workers.

Some have deplored this middle-class intercession in the formulation of trade union purpose, and Selig Perlman and Georges

[4] Hyman and Fryer 'Trade Unions', p. 174.

Sorel, each from a quite different perspective, have spoken for them (See Chapters 2 and 3 above). For the rest, such intercession has been considered at least advisable; and many have thought it absolutely essential. 'Trade unionism . . . is so little sufficient unto itself that the very idea . . . of what it should do had to come to it from outside.'[5] The words are those of Jean Grave, an anarcho-syndicalist, but Lenin was one of many who agreed (see Chapter 4 above). The central point, however, is that whether approved or deplored, the issue of trade union purpose has excercised the minds of outsiders on a scale and to a degree that marks the trade union out among modern non-governmental associations.

The principal reasons for this attention are two fold, and may be briefly stated. The trade union provides a specifically selective organizational framework for directly communicating with (and possibly influencing or even controlling) the mass of the people who comprise the human foundation on which the economic life of modern societies depends. Hence the heavy emphasis commonly placed on an educational function by those who have written about union purpose. Secondly, the trade union also provides at least potential access to a peculiar power resource, the strike weapon, which outsiders have been perennially eager either to tap or to curb. The outcome is an institution which is remarkably versatile, as judged by the purposes that have been thrust on it by its students, its well-wishers, its opponents and, above all, by its would-be manipulators.

[5] Quoted in Guérin, *Anarchism*, p. 81.

PART II

THE FORMS OF TRADE UNIONISM

A THEORY IN COMPARATIVE POLITICS

INTRODUCTION

Trade unionism in its various national forms has been the subject of increasing scrutiny during the second half of the twentieth century. The outcome, in the English language alone, is a very large and rapidly expanding literature. So far, however, it is a literature which is deficient in two crucial respects. First, it lacks a practicable scheme by which the institutional forms of trade unionism may be ordered and categorized on a transnational basis, and in terms which are illuminating as well as systematic. Second, it lacks anything approaching a serious explanatory theory of national trade union movements. This part represents an attempt to fill both these gaps.

The first chapter establishes the distinguishing principle, and develops the categories, for a typology of national trade union movements. Chapters 9 to 13 deal, in turn, with each of the five types so defined—elaborating and illustrating them with particular, but not exclusive, reference to the trade union movements of twenty-seven countries. Chapter 14 explores the question of *why* trade union movements differ in the way they have been shown to differ. Chapter 15 formulates a theory of national trade union movements.

The emphasis in the illustrating cases is predominantly contemporary, although the historical dimension is invariably acknowledged—and sometimes (as in Chapters 11 and 12) bears the burden of the illustration. Precise contemporaneity, of course, is impossible—if only because of my reliance on available published sources, in English, and the rapidity with which circumstances can change. Between two of the drafts of this part, for example, a military coup began a process which put in question my earlier inclusion of Fiji in the select band of countries possessing what I have categorized as 'autonomous' trade union movements. But, in any event, the value of the cases, as illustrations of types or subtypes, does not depend on contemporaneity at all—as the use made of historical illustrations indicates.

The use of the word 'movement' here is not to be interpreted as implying anything of ideological significance about trade unions or trade unionism. In particular, I am not to be taken as following the usage of Selig Perlman, and others, who have held that 'movement presupposes a feeling of solidarity which goes beyond the boundaries of a single trade and extends to other wage earners'.[1] By 'trade union movement', I intend simply to designate all trade unions, as defined in Part I, and all inter-union bodies whose constituents are confined *exclusively* to trade unions (so excluding political parties with union affiliates) within the borders of a single country. Still on terminology: I have preferred the term 'transnational' (though 'cross-national' would serve as well) to 'international' because of the latter's association with trade union organizations that are not confined in their membership by the frontiers of a single country.

Readers of studies in transnational trade unionism are, all too often, drowned in a sea of initials and acronyms. It is with this peril in mind that I have limited these forms of abbreviation to especially pertinent bodies, and otherwise relied on full or shortened verbal versions of organizational names.

[1] Perlman, *A History of Trade Unionism in the United States*, p. 9.

8

OF TRADE UNIONS AND THEIR TYPES

Among the most common forms of association in the world today are bodies which claim the title of 'trade union', 'labour union', 'workers' association', or something recognizably similar—although, quite often (as, for example, in 'Solidarity'), the link may not be immediately obvious to an outsider. Associations of this kind are found under political regimes of every colour, in all the richer nations of the world, and in almost all of the poorer. As (at least ostensibly) non-governmental organizations, they are surpassed in the numerical bulk of their formal adherents only by religious bodies, and not invariably even by them.

Unlike religious bodies, however, trade unions are unequivocally modern institutions, a product of industrialism.[1] Their roots can be discerned clearly in the eighteenth (and perhaps, tentatively, even in the seventeenth) century. Their foundation, in a continuing and stable sense, can be traced back in some cases to the mid-nineteenth century. But they are, above all, creatures of the twentieth. It is the twentieth century that has seen these 'transient' and 'evanescent' associations, as a sympathetic but unpercipient American Cardinal described them in 1887,[2] emerge in force, world-wide, and as institutions which are often thought of as wielding great power—or, at least, as significant repositories of power.

Hugh Clegg, concerned with trade unions on a transnational scale, once wrote of 'the immense variety of trade unionism'.[3] On the other side of the Atlantic a half-century earlier, and with an eye to the narrower context of the United States, Robert Hoxie had less crisply described trade unionism as 'one of the most complex, heterogeneous and protean of modern social phenomena'.[4] Both

[1] Milton and Rose Friedman (*Free to Choose*, pp. 229–31) expressly deny this; but, in tracing the origins of trade unionism back 2,500 years to the disciples of Hippocrates, they have ignored the crucial factor distinguishing the trade union from other types of association—that its members are *employees*.

[2] In Abell, *American Catholic Thought on Social Questions*, p. 159.

[3] Clegg, *Trade Unionism under Collective Bargaining*, p. 1.

[4] Hoxie, *Trade Unionism in the United States*, p. 35.

were right. Nevertheless, Clegg's transnational perspective is the more telling of the two.

It is true that some national trade union movements, like the American, are characterized by considerable internal variety. On the other hand, within the trade union movements of most countries, it is uniformity (or, at least, similarity)—not diversity—which is the rule. The element of variety is in general most pronounced when it comes to comparisons between countries,[5] that is, between different national trade union movements. From a bird's-eye view, they exhibit a bewildering diversity of structure and, particularly, of behaviour.

This diversity begs the question of classification. That is to say, it raises the issue of whether it is possible to introduce order into confusion by categorizing national trade union movements in terms which highlight differences of significance.

Typologies of trade unionism have usually focused on individual union bodies within the framework of a single national trade union movement. Almost invariably, too, they have been formulated with an eye only to Western industrialized societies. Structural classifications have been common in this connection, usually employing such categories as craft, industrial, and general unions, but they have also taken more elaborate forms.[6] H. A. Turner's taxonomy of 'open' and 'closed' unions similarly depends on structural considerations.[7] Apart from structure, the nature of union memberships as a criterion has produced, most notably, classifications turning on the distinction between manual and non-manual employees. Also concerned with union memberships, if from a quite different perspective, is Amitai Etzioni's classification in terms of predominant 'compliance structures', which refers to the means employed by trade unions to control their members.[8] But none of these criteria provides distinctions of obvious value to transnational comparisons.

Purpose, too, has been enlisted as a means of classifying trade unions. It was employed, notably, by Robert Hoxie when he specified a number of 'functional' union types including 'business',

[5] See Millen, *The Political Role of Labor in Developing Countries*, pp. 10–11, on this point.

[6] See esp. Hoxie, *Trade Unionism in the United States*, pp. 38–43.

[7] See Turner, *Trade Union Growth, Structure and Policy*, p. 114.

[8] Etzioni, *A Comparative Analysis of Complex Organizations*, pp. 59–67.

'friendly or uplift', 'revolutionary', 'predatory' and, as an afterthought, 'dependent' (together with a variety of possible sub-types: 'socialistic', 'quasi-anarchistic', 'hold-up', 'guerilla', 'bargaining', 'constructive business').[9] Hoxie's classification was over-elaborate and indecisive, which may perhaps be explained by the fact that his book was compiled posthumously from his lecture notes. It was, in addition, designed to cope with the American trade union scene alone. Nevertheless, the substance of two of Hoxie's categories was subsequently applied to the field of transnational comparison. His 'business' and 'revolutionary' unionism, translated as 'economic' and 'political' unionism, gained wide currency during the period after the Second World War among writers (particularly American) on comparative trade unionism. The American trade union movement was almost invariably taken to represent the purest, existing form of economic unionism. Frequently, moreover, it was assumed to be the only *genuinely* non-political movement—an assumption which effectively implied that there was not a great deal of point to a classification turning on 'economic' and 'political' purposes.

A second two-way classification reliant on purpose is implicit in the depiction of trade unions in Communist states as distinctively 'dual functioning'—that is, as fulfilling the twin purposes of promoting production and protecting workers against managerial excesses.[10] However, this product of Lenin's ingenuity (see Chapter 6), while useful in providing a distinction between received ideologies, is too deeply flawed as a description of empirical reality.[11]

Purpose has also played a part in a more explicit, less limited (because three-category) and notably popular taxonomy. This classified trade union movements by reference to the 'three-world' formula: Western democratic, Communist, and non-Communist 'developing' (originally, 'underdeveloped') countries. The standard outcome was a book devoted mainly to the trade union movements of some Western industrialized societies, but with a number of sidelong glances at the Soviet bloc and a tacked-on chapter or two dealing with some third-world countries. Typically, there was no systematic attempt to establish in general terms any distinctive

[9] Hoxie, *Trade Unionism in the United States*, pp. 45–51.
[10] See Ruble, 'Dual Functioning Trade Unions in the USSR'.
[11] See generally, Pravda and Ruble, *Trade Unions in Communist States*.

connections between these three categories, on the one hand, and trade union organization or behaviour on the other. But what generally came through was a purpose-oriented distinction, at least in part. Thus, versions of 'economic' and 'political' unionism (referring usually to both goals and methods) figured largely in the treatment of Western trade union movements as compared with those of developing countries. In contrast, the dividing line between Western and Communist movements tended to be drawn mainly in terms of a quite different criterion: the presence or absence of state control. On the other hand, these two criteria were more consistently deployed in one recent variation of the three-world formula. Michael Poole distinguished each of three 'types of trade union' in terms of goals ('instrumental', 'political', and 'integrative') and degree of 'independence from state and management'—the types being 'trade unionism under collective bargaining', 'disjunctive-type trade unions' and 'trade unionism under socialism'.[12] But as he applied these three types, the results differed expressly from the standard three-world outcome only in that the trade unions of France and Italy were separated from those of other developed nations and located, instead, in the 'disjunctive-type' category alongside those of the third world.[13]

Generally speaking, however, the principle on which the three-world formula turned (as a classification of trade union movements) had to do directly, not with trade union movements as such, but with the nature of the politico-economic systems which enclosed them. A similar approach figured in a more elaborate classification of national trade union movements which appeared in a study by Kerr, Dunlop, Harbison, and Myers, first published in 1960, under the title of *Industrialism and Industrial Man.* Five types of trade union movement were identified in terms of the nature of a country's 'industrializing élite': the procedure and its results are outlined in the appendix below.

These approaches by way of politico-economic systems fall well short of providing a satisfactory—that is, useful—solution of the classification problem. Thus Kerr and his colleagues were content simply to set out, if quite elaborately,[14] the characteristics they attributed to the labour organizations associated with each type of industrializing élite. They made no attempt to specify even in

[12] Poole, *Industrial Relations*, p. 74.
[13] Ibid. 84. [14] See appendix.

passing, let alone examine systematically, particular trade union movements as illustrations of their types. This throws the reader back to their initial discussion of the five élites, and to the references made there (even then, fairly casually and unsystematically) to specific countries. But the outcome, in terms of trade union movements at least, is some very strange bedfellows. To take a particularly glaring example: it *appears* (whatever Kerr and his co-authors may have intended) that the trade union movements of Japan, Iran, Peru, and West Germany are to be regarded as being of essentially the same type, simply because the same type of industrializing élite (dynastic) was operative in all four cases.[15] Thirty years ago, as well as now, it was on any other criterion a conspicuously motley quartet.

Similarly, those more numerous authors and editors who have adopted the three-world formula have usually been content, when it comes to the third-world countries, to deal with one or two cases which seem to have been selected either at random or primarily on the basis of the specialities of available contributors. Yet, while the similarities within this group of countries are undeniable in some aspects (especially the macro-economic), there are great differences in others; and that is certainly so in relation to trade union organization. The variations within the third-world grouping are usually acknowledged in some degree (at least to the extent of distinguishing Latin America from Africa and Asia on the ground of 'political development'); and sometimes they have been emphasized with particular care by those using this approach.[16] Nevertheless, there has been the same tendency, as in the case of Kerr and his co-authors, to cling to the questionable assumption that this categorization of countries represents a matching and equally justifiable categorization of trade union organization.

The attraction of employing a classification of politico-economic systems (as in the three-world and industrializing-élite cases) to classify trade union movements is obvious enough. Such systems are comparatively easy to identify and distinguish. But the appropriateness of the transference from system to movement depends on the presumption of a necessary and sufficient causal link between the distinguishing systemic characteristics and the character of an associated trade union movement. That presumption, as it happens,

[15] See Kerr *et al.*, *Industrialism and Industrial Man*, pp. 37–9.
[16] See e.g. Kassalow, 'The Comparative Labour Field', pp. 10–13.

does largely hold up in the case of countries with Communist governments (three-world formula) or led by 'revolutionary intellectuals' (industrializing-élite formula).[17] But otherwise, as shown below in Chapter 14, it is subject to heavy qualification. What this points to is the conclusion that it is trade union movements themselves, rather than the political or economic context in which they operate, which must be given priority in any attempt to establish a workable typology of such movements. In other words, the distinguishing criteria of the typology need to be derived directly from the trade unions and whatever aspect of their structure, behaviour, or experience is involved.

Two approaches along these lines were made not long after the publication of *Industrialism and Industrial Man*. One was in Bruce H. Millen's *The Political Role of Labor in Developing Countries* (1963), concerned chiefly with Africa and Asia; the other in *African Trade Unions* (1966) by Ioan Davies. Despite their concern with developing nations, both authors discussed the taxonomic issue with reference to developed nations as well. Millen's treatment of the issue is much the more careful and comprehensive of the two. His book figures in Davies's list of suggestions for 'further reading', but there is little or no sign of it in what Davies had to say on the classificatory issue.

As shown below in the appendix, Millen formulated a 'spectrum' of seven categories, one of them with two sub-divisions; while Davies provided two separate classifications—a simple one distinguishing African from European trade union movements, and a more complex one dividing the movements of 'the developed world' into three categories, one of them with three sub-divisions. Underlying these differences between Millen and Davies, however, there was a broad similarity in their approach to the central matter of distinguishing criteria. Both of them focused on the issue of trade union *autonomy*. In addition (if with considerable differences in emphasis, detail, and outcome), both sought the indicators of such autonomy in trade union relationships with governments and with political parties.[18] On the other hand, neither of them systematically formulated or fully exploited their common insight. Nor was it developed by others.

Under the three-world formula, as we have seen, the autonomy factor effectively operated as a distinguishing principle (in the form

[17] See appendix. [18] See appendix.

of degrees of state control) in relation to Western and Communist trade union movements, but not in relation to those of the non-Communist developing countries. It has been employed, in much the same form and within much the same limits, in other connections—as, for example, by George Bain and Robert Price (1980) when they identified the kind of trade union movements they were concerned with, in a major study of union growth, by drawing a distinction between trade unions that spoke for their members and unions that were dominated by either employers or the state.[19] Jack Hayward (1980) pointed towards the same factor, and implied a more specific interest in the classification issue itself, when he foreshadowed the possibility of being able 'to distinguish between types of trade union movements according to their relationship with their politico-economic environments'.[20] But, in the course of a brief and unsystematic account of ways in which the trade union movements of Western Europe varied, he was content with no more than a sketch of some forms of union–party and union–state relationships.

Gian Cella and Tiziano Treu (1982), on the other hand, pursued the matter more earnestly, as shown below in the appendix. They developed five types ('models' and 'patterns' were their preferred terms)[21] of trade union movements. These were distinguished on the basis of eight 'dimensions'. It is not clear, however, how the five 'models' were derived from the extremely elaborate statement of their 'dimensions'. The confusions which this degree of complexity can give rise to are illustrated, on the one hand, in the uneven character of the accounts which Cella and Treu provided of the five categories in their classification; and, on the other hand, in their apparent uncertainty about whether the categories applied exclusively to national trade union movements as a whole, or as well to *segments* of such movements.[22] At the same time, Cella and Treu echoed Millen and Davies in so far as they included, among their eight 'dimensions', one involving trade union relations with political parties and another involving union relations with the state—and, to this extent, incorporated the autonomy factor in their scheme.

[19] Bain and Price, *Profiles of Union Growth*, p. 2.
[20] Hayward, *Trade Unions and Politics in Western Europe*, p. 2.
[21] Cella and Treu, 'National Trade Union Movements', pp. 221–2.
[22] See appendix.

A TYPOLOGY OF TRADE UNION MOVEMENTS

A workable classificatory scheme requires a distinguishing principle which has three characteristics. The first is that is reasonably *clear-cut*—in other words, not too complicated. The second is that it is readily *observable*, which in the present case means that its indicators are discernible without too much guesswork in a huge secondary literature that is highly variable in quality and in focus.[23] These two characteristics are methodologically critical, in that they supply a principle which is capable of being applied systematically and with a minimum of confusion.

The third characteristic is that the distinguishing principle is *relevant* in the sense that it yields categories which are demonstrably significant. This is the supremely critical consideration—an assertion which may be best supported by an illustration. Political parties have been the subject of intense classificatory efforts by students of political systems since the Second World War, and especially since Maurice Duverger's pioneering work on the topic.[24] The simple number of parties in a political system has been most used, and abused, as a taxonomic principle; and it remains not only 'a highly visible element' but also, despite its critics, a highly relevant indicator of the nature of modern political systems.[25] The same principle, however, has quite different implications in the trade union case, where confederations (defined below in this chapter) constitute the nearest counterparts of parties. Clearly, in this case, the simple numerical principle admirably satisfies the two methodological requirements (clear-cut and readily observable). But as to its relevance—who, in their right mind, would equate the British and the Soviet trade union movements simply on the ground that both are effectively embodied in a single confederation?

The autonomy factor stands up to these three tests. Loosely formulated and casually applied though it was in the work of Millen and others, it is nevertheless capable of providing a viable distinguishing principle for a classification of national trade union movements.

First of all (to reverse the earlier order), as to relevance, there can be no question of the importance generally attached to the relations

[23] See the introductory remarks to the Bibliography to Part II below.
[24] *Political Parties*.
[25] Sartori, *Parties and Party Systems*, p. 121.

between trade unions, on the one hand, and the state and political parties, on the other.

Secondly, on the issue of observability, trade union relations with the state and political parties have a particularly high profile in the literature—and, for example, a consistently higher profile in general than trade union relations with employers/managers.

Finally, on the question of clarity, the implications of the autonomy factor can be systematically formulated, with an eye to states and parties, as follows. There are notionally three distinctive positions which trade unions may occupy in relation either to a state or to a political party. In the first place, they may be dominated by it. In the second place, they may dominate it. In the third place, they may neither dominate nor be dominated. The first position I shall call the *ancillary* position; the second, the *surrogate* position; and the third, the *autonomous* position.

Thus a trade union movement may be said to be in an ancillary position when major union bodies are characteristically subordinate or subservient to either a political party, the state, or both.

In the surrogate position, it is the trade union bodies that dominate the party or the state—a domination that may ultimately take the form of union bodies acting in place of party or state by displaying qualities, espousing purposes, and discharging functions normally associated with parties or with states.

The autonomous position does not necessarily imply complete independence of parties and state. It is better seen, rather, as being intermediate between the ancillary and surrogate positions—in the sense that the trade unions are neither clearly subservient to, nor clearly dominant over, either party or state.

These three positions, given that there are three nominally distinct organizations involved, yield five categories:

1. party-ancillary;
2. state-ancillary;
3. party-surrogate;
4. state-surrogate;
5. autonomous.

Most national trade union movements, as it happens, fall into one or other of the two ancillary categories. Autonomous trade union movements are less abundant. Those of the surrogate type are rare—indeed, if it were not for the peculiar history of one trade union movement, the state-surrogate category would be mentioned (but no more) only for the sake of logical symmetry.

State-ancillary is the single most populous category. The trade union movements which figure in it are mainly those in authoritarian political systems. In such systems, the trade unions are often controlled through the apparatus of the ruling political party. Cases of this kind are here located in the state-ancillary (rather than the party-ancillary) category on the ground that there is no politically significant distinction between the party and the state in what has been appropriately labelled a 'party-state system'.[26] It follows that the *party*-ancillary category is reserved for cases in which a dominating party is either not a ruling party or, if in government, operates in the context of a genuine competitive party system which holds the possibility that the party may lose power by peaceful means.

All categories are to be interpreted broadly, allowing for considerable variations in detail between cases. In other words, like most classifications in the socio-political area, it is a matter of approximation and of loose rather than precise fits—which may involve rough justice at the margins. In addition, there is the historical factor to be taken into account. Over time, trade union movements may shift (and very often have) from one category to another.

On the empirical side, two preliminary points need to be made. The first is that, so far as trade union organization is concerned, the focus is on *confederations*—that is to say, on peak inter-industrial bodies (often officially entitled congresses, councils, or federations, as well as confederations) which are national in scope, at least in ambition, and incorporate individual trade unions or sectional federations by way of formal affiliation, or some tighter linkage. As a structural form, confederations are not only universal (for all practical purposes) in the modern world, but are by far the most accessible form in terms of the literature on trade unions. The second, and more important, preliminary point is that the emphasis in the illustrating cases presented in the chapters that follow is firmly on effective relationships, which are rarely a mirror-image of the formal relationships set out in constitutions and other official documents. Formal relationships invariably fail to tell the whole story, and all too often tell the wrong story.

[26] Sartori, *Parties and Party Systems*, p. 44.

9

PARTY-ANCILLARY MOVEMENTS

Trade union movements of the party-ancillary type invariably contain two or more national confederations which are conventionally distinguished from each other (by themselves and by outsiders) in terms that are predominantly political, rather than industrial or occupational. In some cases these distinctions may be given an expressly religious dimension, and occasionally a racial or communal dimension.

The party-ancillary movements of Western Europe exist in a belt of countries running from the Netherlands, Belgium, and Luxembourg in the north, through France and Switzerland to Italy, Spain, and Portugal in the south. They were once quite common in Africa, during colonial times, but the type appears to have survived only in South Africa. Their Asian counterparts weathered independence with more success, and persist in India, Pakistan, Bangladesh, Sri Lanka, Thailand, and the Philippines. The party-ancillary type is also prominent in the Americas, figuring in at least six countries of South America (Colombia, Ecuador, Peru, Venezuela, Belize, and Guyana), possibly five in Central America (Costa Rica, Dominican Republic, Guatemala, Honduras, and Panama), and at least four in the Caribbean (Jamaica, Trinidad, Antigua, and Puerto Rico).

In all party-ancillary cases, the lines of division between trade union confederations reflect a tendency on the part of the confederations to identify themselves with specific political parties. Those that are not in some way linked to a particular party invariably define themselves, in the first instance, with reference to others that are. Quite often, confederations (or their direct antecedents) were originally the creation of an associated political party (or its antecedent). But in no party-ancillary case nowadays, outside Latin America (and, there, probably only in Peru),[1] does the linkage between them appear to involve the kind of organic constitutional connection signified by the formal affiliation to the party of the confederation or its constituent trade unions. The

[1] See Johnson, *The Political, Economic and Labor Climate in Peru*, pp. 112, 215.

nearest thing to a formal connection, and it is quite rare, is a standing consultative body on which both party and confederation are officially represented. Otherwise, informal connections are the rule—involving, above all, overlapping memberships in the case of leading party and trade union organs.

The relationships existing between confederations and associated parties differ within and among particular trade union movements. There are, especially, great variations in the degree to which different parties dominate their associated confederation in the sense of exercising actual organizational control over it. The range is from complete control (in the literal sense of 'ancillary') to none at all. In the latter case the confederation is operationally independent. Nevertheless, it is still to be regarded as ancillary to the party—and in a sense dominated by it—to the extent that it relies on the party for the political definition which distinguishes it from other confederations. This minimal form of party domination may be said to operate also (if in a more generalized sense) in the case of confederations which, by disowning *all* political parties, similarly distinguish themselves from other confederations in party-political terms. But there is one circumstance in which a confederation, even though defining itself with reference to an associated party, cannot be regarded as even minimally dominated by the party, or in any real sense ancillary to it. This is when it is the confederation which exercises effective and continuing organizational control over the party: that circumstance is considered in Chapter 11.

CONFEDERAL COMPETITION

A central issue, in the case of party-ancillary movements, is the nature and degree of the competition which political differences impose on the trade union confederations involved. The essential point to be made in this connection has to do with the context of such competition. Although the structural fragmentation of party-ancillary movements is defined in political terms, the confederations through which it is expressed purport to operate at the industrial level as well. At that level, they are not merely rivals because of conflicting political allegiances and aspirations, they are organizational duplicates in terms of their prospective membership and professed industrial concerns. Their formal constitutions put them in head-on competition with each other. In principle, that is to say, through their constituent trade unions they compete for the allegiance of the same employees and for the recognition of the same employers.

In practice, the character of the competition varies from case to case, both within a particular trade union movement and, more generally, as between different movements. The spectrum is wide. At one extreme, there is the kind of competition which can flare into beatings and killings. At the other extreme, there is the kind which is all but buried under confederal and/or party co-operation. This spectrum is explored below by way of five cases, running from intense to mild confederal competition. The cases, in that order, are India, France, Italy, Belgium, and Venezuela.

India

The first national trade union confederation was formed in 1920 in response to requirements which the newly-created International Labour Organization had laid down for governments wishing to be represented at ILO conferences. The colonial link with Britain was evident in the naming of the All-India Trade Union Congress (All-India TUC). The parties quickly moved in. Communist union leaders had won the battle for control by the end of the decade. Unions led by members of the Indian National Congress broke away to form their own confederation; and a second split, this time among the Communists themselves, produced a third confederation shortly afterwards. The struggle for independence had by 1940 brought all these groupings together again in the All-India TUC, but not for long. Differences over attitudes to the war produced new splits. The All-India TUC survived, by now under firm Communist control, into the post-war years and the achievement of national independence, but faced mounting competition. In 1947, shortly before independence, the Indian National Trade Union Congress (Indian National TUC) was set up under the auspices of the Indian National Congress, and quickly displaced the All-India TUC by winning official government recognition as the 'most representative' labour organization. The following year saw the emergence of the Indian Labour Association (HMS—Hind Mazdoor Sabha), dominated by members of the Socialist party. In 1949, representatives of a collection of extreme left parties, including the

Revolutionary Socialist party, founded the United Trade Union Congress (United TUC).

By 1949 there were thus four national confederations, each of them identified with a specific party or group of parties. This pattern of political fragmentation was reinforced by an intervention from the right wing in 1955, when the National Labour Organization (BMS—Bharatiya Mazdoor Sangh) was founded as the self-professed 'labour front of the Jana Sangh' party.[2] Fifteen years later, the left wing was to add a further confederation to the list when the Communist Party of India (Marxist) threw up the Centre of Indian Trade Unions as a breakaway from the All-India TUC and the Communist Party of India. Another left-wing group, the Marxist party known as Forward Bloc, unable to re-enter the All-India TUC on satisfactory terms, created yet another confederation in 1971, the Trade Union Co-ordination Committee.

In the mid-1960s, it had been possible to cite the Swatantra party, on the conservative right, as the only substantial Indian party that 'does not sponsor a labor federation'.[3] But even the Swatantra party, in 1968, formally decided that it had 'to enter into the trade union field if only to "depoliticize" it'[4]—although, as it happened, the party was electorally obliterated before it could carry out this resolve. By the early 1980s, according to allegedly dubious official figures,[5] the largest single confederation was still the Indian National TUC, containing 38 per cent of all union members covered by ten national confederations. Its nearest competitor, in size, was the Jana Sangh's BMS, to its right, followed by the Socialist party's HMS, to its left—and thereafter, further to the left, a long tail of at least seven confederations headed by the All-India TUC, the United TUC, and the Centre of Indian Trade Unions.

In no case is there a formal nexus, in the shape of official affiliation, between a party and a trade union organization. The closest thing to such a nexus is the Co-ordination Committee set up in 1958 between the Indian National TUC and the Congress party. Nevertheless, the reality is that the links forged between party and confederation, by way of overlapping memberships at the leadership level, have meant that party-political considerations have tended to carry greatest weight in determining the behaviour of each confed-

[2] Baxter, *The Jana Sangh*, p. 188. [3] Ibid.

[4] Chatterji, *Unions, Politics and the State*, pp. 145–6.

[5] See Roy, 'Charade of Verification', pp. 1818–20.

eration. The confederations, in other words, are all effectively dominated by parties, for whom unions have a peculiar interest in that they provide a means of communicating with 'the élite of the country's labour, the industrial workers'.[6] But party control of an associated trade union confederation tends to be much tighter and more complete on the extreme left, and specifically in the case of the Marxist-oriented parties which have relatively little support in general electoral terms.

The outcome is a structurally fragmented trade union movement in which, at the level of the shop floor, unions are identified by rank-and-file workers primarily in terms of their party allegiance. Another consequence, at the same level, is an intense competition between unions—a competition which often overflows into violence. At times the competition may be muted as an outcome of party alliances. Thus a split in the Congress party produced an alliance between it and the Communist party of India, which in turn (by early 1972) resulted in the Indian National TUC, the All-India TUC and the HMS forming a joint National Council of Central Trade Unions. The confederations further to the left reacted by also coming together, later in the same year, to form the United Council of Trade Unions. These co-operative arrangements on the trade union side came to an end when the parties divided once again.

France

French trade unionism is renowned for a long-standing and influential syndicalist tradition involving, in particular, a distrust of political parties as an instrument of working-class action. It was a tradition born during the 1880s and 1890s out of the faction-ridden politics of the Socialist political movement (which could claim five separate national parties by 1899) and the emergence of Anarcho-syndicalism as an initially French variant of anarchistic thought. But the tradition, while of undeniable significance, has not been sufficient to deprive parties of a weighty influence in the French trade union movement.

The General Confederation of Labour (CGT), founded in one form in 1895, has remained the major confederation ever since. It was formed independently of parties; and in 1906, by way of the 'Charter of Amiens', it clinched the point by formally nailing the

[6] Raman, *Political Involvement of India's Trade Unions*, p. 39.

standard of revolutionary syndicalism to its masthead. But, as elsewhere, with the onset of the First World War the CGT denied its professed ideology by collaborating both with France's war effort and with a unified Socialist party. It continued this policy of party collaboration after the war; and so did a new United General Confederation of Labour (United CGT) formed in 1921 by union leaders who broke from the CGT in order to align with the Communist party. It was left to the Catholic trade unions to preserve the non-party element of the syndicalist tradition in the shape of the French Confederation of Christian Workers (Christian Confederation), formed in 1919.

The twists and turns of Soviet foreign policy during the 1930s saw the constituents of the United CGT back in the CGT by 1936, and out again in 1939. The CGT and the Christian Confederation continued as co-operating underground organizations throughout the war. After the Communists joined the resistance movement, following the German invasion of Russia, their union wing was once again formally admitted to the CGT. With the end of the war, it soon became clear that the balance of power within the CGT had shifted to the Communists. The minority, mainly Socialist, faction broke away in 1948 to form a new confederation, the CGT-Workers' Movement (CGT-FO). There has been no reconciliation in this case.

One other major split provides a further demonstration of French trade unionism's talent for splintering along political lines. From 1944 to 1953, the bulk of the Christian Confederation's leadership was associated with a Catholic party, the Popular Republican Movement; but the relationship posed difficulties because of a growing anti-clerical element in the Christian Confederation's membership. In 1964 this element finally triumphed. The word 'Christian' was dropped from the confederation's title and it was renamed the French Democratic Confederation of Labour (Democratic Confederation). The split that followed produced another, much smaller confederation which retained the old name with the addition of *Maintenu*. Since then, the Democratic Confederation has pursued a notably erratic course. It entered into a formal alliance with the CGT, involving both industrial and political co-operation; and then, after the worker–student demonstrations of May 1968, moved leftward of the CGT into an effectively syndicalist position and a more militant approach generally. A strengthened association with the Socialist party subsequently took it back into a formal partnership with the CGT, through the *Union de la Gauche* constructed between the Communist, Socialist, and Left-Radical parties in the early 1970s. The breakup of this arrangement, after it failed the test of the 1978 general election, saw the Democratic Confederation again affirm its independence of all party ties while, this time, moving to the right of the CGT with limited policy aims and a new emphasis on industrial negotiation rather than strikes.

The volatility of the Democratic Confederation illustrates the strength within French trade unionism of the anti-party element in the syndicalist tradition. So, too, does the way in which both the CGT-FO and the Christian Confederation (*Maintenu*) have maintained their distance from the parties since the 1960s. And the same policy has been followed by a fifth confederation concerned exclusively with technical, professional, and supervisory staff (General Confederation of Cadres) since its foundation in 1944. Nevertheless, in the end, it is precisely the *rejection* of that tradition by the CGT which has played the critical part in shaping the character of the French trade union movement.

The CGT, as we have seen, was under Communist control by the end of the Second World War. Ever since, its relationship with the French Communist party has been unequivocally one-sided. 'The CGT *never* adopts an autonomous position with regard to the [party]; on the contrary it always follows the Party line (more or less strictly depending on the circumstances).'[7]

The CGT's subservience to the party is critical because the CGT is the lodestar of the French trade union movement. It is the largest and most comprehensive of the confederations, covering perhaps as much as 40 per cent of those union members claimed by all five confederations, though their precise memberships are uncertain. Its prominence, arising from its size and coverage, mean that workers very often join CGT unions for reasons which have little or nothing to do with its political alignment. In contrast, 'the employee who belongs to the CFDT [Democratic Confederation] or the CGT-FO is [usually] someone who wants to belong to a union, but who, for moral or political reasons, *does not want to belong to the CGT*'.[8] This reactive pattern is evident at the leadership level as well, in that

[7] Lavau, 'The Changing Relations between Trade Unions and Working-Class Parties in France', p. 440.

[8] Ibid. 439 n.

the other confederations and their unions 'define themselves', above all, in relation to the CGT.[9] And, with exceptions from time to time, they prefer to compete rather than co-operate with each other in relation to membership recruitment and employer recognition.

Italy

In 1906, the year that the French CGT adopted the Charter of Amiens, trade unions associated with the Italian Socialist party formed their own General Confederation of Labour. A second confederation, inspired directly by French syndicalism, emerged six years later in the shape of the Italian Syndical Union. Two more were formed in 1918. The larger, the Italian Confederation of Workers, was associated with a Catholic political party. The other, the Italian Labour Union, had affinities with Fascism. None of these four confederations survived Mussolini's drive, once he gained power in 1922, to destroy all trade union organizations other than those directly controlled by the Fascist state.

In 1944, with the end of the war in sight, representatives of what were to become the three major Italian parties (Socialist, Communist, and Christian Democrat) negotiated the formation of a single trade union confederation, the Italian General Confederation of Labour (CGIL). The agreement provided that the CGIL was to be officially independent of political parties; it also provided that each of the three parties was to be equally represented in all key executive positions and bodies—which meant, for example, that the CGIL had three general secretaries. This arrangement endured while the three parties co-operated in the government of Italy, but was doomed once the Christian Democrats excluded the Communists and the main Socialist party from the ruling coalition in mid-1947. The Christian Democrat union leaders were the first to break away from the CGIL: they formed the Free Italian General Confederation of Workers in 1948. The next to go, in 1949, were union leaders linked to other parties in the coalition with the Christian Democrats, the Republican party and a right-wing Socialist party: they set up the Italian Labour Federation. The following year, however, the leading members of these two

[9] Lavau, 'The Changing Relations between Trade Unions and Working Class Parties in France', p. 439; and see also Gallie, *Social Inequality and Class Radicalism in France and Britain*, pp. 164–5.

breakaways joined forces to form a single confederation, the Italian Confederation of Workers' Unions which was to last. The merger left in the cold a strongly anti-clerical group within the Labour Federation who refused to go along with the merger. They, in turn, found new partners in yet another CGIL breakaway which had taken out most of the remaining Socialist leaders. The outcome of this new alliance was the Italian Workers' Union.

By 1950 there were thus three rival confederations (this is to omit a Fascist-oriented confederation, founded in that year, which has persisted but by all accounts has remained negligible). The CGIL, the oldest and largest, was dominated by the Communist party, but included a segment identified with the main Socialist party. The Confederation of Workers' Unions consisted predominantly of Christian Democrats with a leavening of right-wing Socialists. Finally, there was the smaller Italian Workers' Union, identifying with the Republican party and right-wing Socialists. At this stage, and up until the later 1960s, there is no doubt that each confederation was effectively subordinate to the party or parties with which it was associated through its leadership. This was, as it has been described, a period of 'party tutelage'[10] for the confederations, and major trade union positions were 'fiefs at the disposal of political parties'.[11] There was, however, a shift in this pattern from the end of the 1960s, in that the confederations moved apart from the parties. One aspect of this shift was the extent to which they co-operated instead of competed industrially, a process that had in fact begun a decade earlier.

During the 1950s the fragmentation of the union movement had, as in France, militated against the development of collective bargaining. Then, in the early 1960s, individual unions from the three confederations began to co-operate in industrial negotiations, first at the regional and local levels, and later nationally—a development which seems to have owed a great deal to a new generation of activists in the Confederation of Workers' Unions. The trend to closer working relations was especially pronounced in the metal industry; and it was dramatically demonstrated in the events and the successes of the 'hot autumn' of 1969.

By the end of the sixties, many unions in the three confederations were thus tied into joint bargaining arrangements. This deviation

[10] Farneti, 'The Troubled Partnership: Trade Unions and Working-Class Parties in Italy', p. 422.

[11] Kendall, *The Labour Movement in Europe*, p. 171.

from the Indo-French competitive model involved two further deviations. One was a stronger orientation within the confederations towards industrial, as distinct from political, concerns. The other was a strengthening of individual union leaderships at the expense of the confederations, in many cases denying the latter the control they had formerly exercised over constituent unions.

Alongside these developments in the *industrial* field among their associated trade unions, the confederations themselves also proved ready by the later sixties to co-operate in the *political* field. Thus, in 1967 the Confederation of Workers' Unions, for the first time in its history, not only joined with the CGIL to present a common set of pensions reform proposals to the government, but took part in a supporting national strike alongside the CGIL. Joint action of this kind, on the part of all three confederations, has since ceased to be so unusual. On the other hand, moves formally to merge them failed, in the early 1970s, as also did a similar attempt confined to the three unions of metal-workers. However, a looser arrangement, involving a joint secretariat in which the three confederations were equally represented, lasted for just over a decade until its collapse in 1984. None of this, on the other hand, obviated competition for membership. The CGIL remains the largest in this respect, with about 47 per cent of the total number of members (excluding retired employees who retain their affiliation, and are usually included in Italian membership figures) claimed by the three confederations.

As for the parties, both the CGIL and Confederation of Workers' Unions congresses in 1969 forced reluctant leaderships to accept the principle of 'formal autonomy from the political parties',[12] by debarring trade union leaders from holding political office, either as parliamentarians or as party officials. But the linkages remain. And to at least one observer, 'it is hard to foresee Italian trade unions becoming emancipated from political parties', in the sense of achieving 'complete independence'.[13]

Belgium

The formation of the Belgian Workers' party in 1885 preceded the French General Confederation of Labour by a decade, the Italian by

[12] Weitz, 'Labor and Politics in a Divided Movement: The Italian Case', p. 237.
[13] Farneti, 'The Troubled Partnership', p. 420.

two, and the All-India Trade Union Congress by well over thirty years. Unlike these specifically trade union bodies, the Belgian Workers' party initially functioned not only as a political party but also as a trade union confederation. Trade unions, along with co-operative societies, friendly benefit societies, and socialist groups, were linked to it by way of formal affiliation. In 1898 the party took the first step towards a separation of roles. It established a Trade Union Commission, to act in the confederal role, with half of its executive members nominated by the party leadership, and the other half by the affiliated unions. Eight years later, in order to attract a number of unaffiliated unions, the commission was reconstituted so that all but two of its executive membership came from unions affiliated directly to it. The two reserved seats were for party representatives; and, in exchange, two seats on the party's executive body were allocated to the Trade Union Commission. It was not until 1937 that the unions affiliated to the Trade Union Commission formally asserted their independence of the party and once more reconstituted the commission: this time as the General Confederation of Belgian Labour. In the meantime, Catholic trade unions, associated with the Catholic party, had formed their own General Confederation of Christian and Free Belgian Unions in 1909. A third grouping, linked to the Liberal party (an anti-clerical conservative party), formed the General Centre of Liberal Unions before the Second World War.

The organizational unity of the left-wing trade unions in Belgium stands in stark contrast to the inter-war performance of their Indian, French, and Italian counterparts. The comparatively small influence of Communism in the Belgian trade unions largely explains this difference. Nevertheless, once Germany invaded Russia, Communists played a major part in the underground labour movement associated with the Resistance. When the war ended, there were three confederal trade union bodies on the left, one of them Communist-controlled; but by mid-1945 all had merged into the marginally renamed General Federation of Belgian Labour (FGTB). Communists played a prominent role in the FGTB until 1948, but their influence in Belgian trade unionism plummeted following their Czechoslovakian comrades' successful coup.

The confederal scene after the war, despite some different names, was essentially the same as before; and it was to endure. The FGTB, associated with the Socialist party, confronted the Confederation of

Christian Unions (Christian Confederation), associated with the Christian Social party; and, trailing the field, the General Centre of Liberal Unions (Liberal Centre) was linked with the Liberal party, subsequently renamed the Party of Liberty and Progress. The Liberal Centre appears to be rather more dependent on its associated party than either the Christian Confederation or the FGTB. On the other hand, the latter two not only identify themselves in terms of the party but, in each case, maintain a formal connection with it by way of representation on a joint consultative body. The FGTB has perhaps demonstrated the strongest sense of independence in its 1964 decision to make union office and a parliamentary seat mutually exclusive. Nevertheless, the party links of both the major confederations have been described as 'close and intimate';[14] and another observer depicted them as involving 'relationships of affinity and general reliance' on the respective parties.[15]

Broadly speaking, the pattern of structural fragmentation is similar to the Indian, French, and Italian cases, but with two striking differences. First, there is no Communist-linked confederation. Second, the biggest of the confederations is not the Socialist-aligned FGTB, but the confessional Christian Confederation which caught up with the FGTB's membership in 1959 and now covers almost half of the total confederal membership.

The Christian Confederation and the FGTB are in direct competition for members over a wide range of industries and occupations. The Liberal Centre covers mainly some white-collar employees. The Christian Confederation has a clear-cut advantage in the Flemish north (from which it draws over 80 per cent of its membership), the FGTB in the Walloon south and in mainly francophone Brussels. The competition for members, once often bitter, is still genuine enough. Nevertheless, the Christian Confederation and the FGTB are accustomed, particularly since the 1950s, to close co-operative working at the industrial level. They and their respective unions usually present a joint case in collective bargaining negotiations with employers, and take industrial action on a joint basis. Political strikes, however, are another matter.

The co-operative relationship between the two major confederations has its counterpart at the parliamentary level where the

[14] Kendall, *The Labour Movement in Europe*', p. 241.
[15] Lorwin, 'Labor Unions and Political Parties in Belgium', p. 260.

Socialist party and the Christian Social party have often been partners in coalition governments. The relationship is echoed in their joint representation, along with the Liberal Centre, on high-level tripartite consultative bodies established by government, the senior of which (the Central Economic Council and the National Council of Labour) date back to the 1940s. As well, the three confederations are all involved in various ways in a diverse array of administrative and advisory bodies ranging across the economic, industrial, and welfare areas. The heads of the Christian Confederation and the FGTB are normally among those consulted by the King when cabinets are in crisis—which, as in Italy, often happens in Belgium, but with quite different implications for trade union involvement because the unions' constitutional status (in a word, their respectability) is less ambiguous.

Venezuela

General Juan Vicente Gomez, in the course of a long (1908–35) and brutal dictatorship, destroyed key traditional elements in Venezuelan politics—the regional *caudillo*, as a source of military power, and the two major parties based on commercial and land-owning interests. Under the somewhat less repressive regime of his immediate successors, a new set of political institutions, including a trade union movement, showed signs of emerging. A *coup d'état* in 1945 broke the pattern of Venezuelan politics even more radically by replacing a military dictatorship with a civilian government based on popular election. A second coup in 1948 turned the clock back. A military junta was followed, in 1952, by the dictatorship of General Marcos Pérez Jiménez. But a third coup in 1958 saw the restoration of civilian government and popular elections. This time they were to last.

The new regime passed one important test six years later. For the first time in Venezuelan history, an elected president gave way to an elected successor. But both men, on that occasion, were from the one party, Democratic Action. The greater test was passed in 1969 when the outgoing president handed over the reins of office to a member of a different party, the Social Christian party. Since then, the presidency has been peacefully transferred between these parties in 1974, 1979, and again in 1984.

Democratic Action (AD) emerged in clandestine form shortly

after the demise of Gomez, and was officially founded in 1941. Its leaders were closely involved with the junior military officers who staged the 1945 coup. It dominated the political scene following the coup, winning the presidency and an overwhelming majority of national congress seats. The period, which ended when its erstwhile collaborators in the army turned against it, has come to be known as 'the AD *trienio*'. The party was banned in 1949; and remained so until Jiménez fell in 1958.

The Social Christian party (COPEI), AD's main competitor, was formally founded in 1946—although its roots stretch back as far as AD's. A second major AD opponent, the Democratic Republican Movement, was one of a number of parties to emerge during the *trienio*. Their much slighter popular support than AD saved them all from suppression until the last months of Jiménez's dictatorship, when COPEI incurred his displeasure. The Communist party, which could trace its origins back to 1931 and the Gomez era, carried little weight.

Many other parties have surfaced since the 1958 coup, the more notable of which have been left-wing breakaways from AD. Nevertheless, AD and COPEI have remained the crucial components of the Venezuelan party system. Together with a third party, the Democratic Republican Movement, they formed an early coalition in the national congress; and the Movement's defection left the AD–COPEI alliance to underpin the first AD presidency. COPEI withdrew from this alliance in 1964. Since then, Venezuela has been exposed to a party system in which AD has not only usually lacked an assured majority in the national congress, but from time to time has had to concede the most powerful office of all, the presidency, to COPEI.

Trade unionism in Venezuela is essentially a product of party organization. As such, it emerged in strength only with the passing of the Gomez dictatorship. The Communists, initially dominant in the area, had their grip broken by AD in a mysterious episode involving unexpected and unexplained government intervention in 1944.[16] Subsequently, during the *trienio*, AD massively extended its organizational influence and, in 1947, was behind the formation of the Confederation of Venezuelan Workers (Workers' Confederation). COPEI, in the same year, entered the field by forming the

[16] See Martz, *Acción Democratica: Evolution of a Modern Political Party in Venezuela*, pp. 257–8.

Copeyano Labour Front (Labour Front) as a party outrider with the function of organizing unions of workers who were COPEI members or sympathizers. During the ensuing decade of military government, the Workers' Confederation was banned outright and the Labour Front functioned under great difficulties.

Following the 1958 coup (as in Italy when the Second World War was ending), the main Venezuelan parties negotiated a settlement of the trade union question. They formed a National United Trade Union Committee consisting of six AD representatives, two from COPEI, two from the Democratic Republican Movement and one each from the Communist party and a non-party group. All parallel local unions organized under the aegis of the different parties were merged; and the four parties submitted a joint slate of candidates (which effectively guaranteed success) in all union elections that year.

The Workers' Confederation, too, was reconstituted on a party-co-operative basis; and though COPEI's Labour Front remained in existence, its activities have since been confined to working within the structure of the Workers' Confederation and its unions. AD, COPEI, and the Democratic Republican Movement (the Communist party joined opposing party groups at this point) presented a joint slate of candidates at the Workers' Confederation congress of 1959, and won the major positions. Since then, AD and COPEI members have run a joint ticket of their own at Workers' Confederation elections, and have throughout retained control of its main offices and a majority on its main organs. The support of COPEI, although invariably the junior partner in terms of its voting strength in the Workers' Confederation congress, has usually been imperative for the maintenance of the AD position; and this has been acknowledged in the offices within the confederation which have been conceded to COPEI members.

By the mid-1960s, the Workers' Confederation was flanked by two others. To the left, there was the United Workers' Confederation of Venezuela (United Confederation) formed by a group consisting of union leaders from the Communist party, some from the Democratic Republican Movement, and some from an AD faction—all of whom had been expelled from the Workers' Confederation in 1961. To the right, there was the smaller Confederation of Independent Trade Unions (Independent Confederation). Like the older Labour Front, the Independent Confederation is an offshoot of COPEI, and was founded (in one form) in 1958. At that time, the party leadership decided that while the Labour Front would work exclusively within the reconstituted Workers' Confederation, nevertheless individual COPEI members should be free to organize separately under the umbrella of the Independent Confederation in order to 'build the Venezuelan Christian labor movement'.[17]

Of the three confederations, the Workers' Confederation is by far the biggest. It covers some 90 per cent of the organized labour force, but more than half of its affiliated membership (of over one million) are peasants and rural workers organized in the Venezuelan Peasants' Federation. The United Confederation and the Independent Confederation, between them, have probably under 50,000 members, but all are non-rural employees. The Workers' Confederation is the one with which governments are accustomed to deal; and the great bulk of its central income is derived from government sources, members' subscriptions being soaked up by its local unions and their federations.

The linkages between political parties and Venezuelan trade unions, at all levels, are close and continuing, although unions are legally debarred from formal party affiliations. Most major union leaders in all three confederations are also major figures in the parties they belong to; and that applies at lower organizational levels as well. Workers' Confederation officials, in particular, frequently represent their party in the country's national congress. Within the Worker's Confederation, its local unions and their federations, the factions of the different parties are in varying measure controlled by a parallel party organization operating at all levels and concerned specifically with the operations of the faction. In the case of COPEI, as we have seen, the Labour Front is the parallel party organization. Elections at all levels of the Workers' Confederation are the subject of party competition, each faction on its own or, more usually, in combination with another or others, presenting a slate of candidates. Seats on union organs are distributed on the basis of proportional representation.

Since 1958, given the Workers' Confederation's domination of the trade union scene, co-operation has been the pre-eminent feature of organizational relationships within the Venezuelan union movement. In its *formal* inclusion of all major political parties, the

[17] Herman, *Christian Democracy in Venezuela*, p. 78.

Workers' Confederation resembles the Israeli and the Austrian cases which are considered in Chapters 12 and 13, respectively. In Israel and Austria the party-political competitive element within the trade union movement is not merely embodied within the one confederation but, more unusually, it is formalized. However, the Venezuelan case differs in that the competition, so far as it relates to confederal organization, is not entirely contained in this way—as the presence of the United Confederation and the Independent Confederation testifies. (A close, if historical, parallel to the Workers' Confederation in this respect is to be found in Chile before the Pinochet coup of 1973.)[18]

The unusual mix of co-operation and competition achieved in the Venezuelan case is illustrated by the open association of COPEI, as one of the two major parties, with two distinct trade union-oriented organizations: the Labour Front, with the function of mobilizing and co-ordinating the party's effort within the Workers' Confederation; and the Independent Confederation with the function of *competing* with the Workers' Confederation. Moreover, the Independent Confederation's competitive role has two particularly notable aspects. In the first place, there is the competition at the level of industrial organization. In the generality of Latin American countries with politically distinguished trade union confederations, it appears that confederal competition does not penetrate below the national and regional levels because rival unionism is usually prevented by law at the level of the enterprise or the plant. Venezuelan law, like that of other Latin American countries, specifies that legal representation of a group of workers depends upon formal recognition by the Ministry of Labour. Nevertheless, rival unions are still apparently found at both the local and industry levels in Venezuela. Thus the Independent Confederation competes with the Workers' Confederation at the local as well as the national level—and this can mean competition with Workers' Confederation local unions which are controlled by Labour Front (that is, COPEI) factions. The United Confederation, the third confederation, is similarly involved in competition with Workers' Confederation unions. At the same time, it is clear that such competition is rarer in Venezuela, and its scale much more limited, than even in the Belgian case.

[18] See Angell, *Politics and the Labor Movement in Chile.*

In the second place, a more subtle aspect of confederal competition in Venezuela is revealed in the observation that 'many unions whose sympathies lie with the other two confederations [Independent Confederation and United Confederation] choose to operate instead within [the Workers' Confederation's] framework'.[19] They do so for pragmatic reasons: the Workers' Confederation is more effective as 'a vehicle for political and bargaining action' than its competitors.[20] But that does not displace ideological loyalties—which means, for example, that the Independent Confederation 'plays its most important role as the voice of the COPEI trade union leaders, whether or not they officially belong to the confederation'.[21] In other words, the competitive element that the Independent Confederation provides (and the point appears to apply equally to the United Confederation), while of little moment in an organizational sense, is highly significant when it comes to the articulation of ideological differences within the trade union movement.

COMPETITION, RELIGION, AND RACE

There would seem to be some connection between degrees of party control, the nature of inter-party relationships, and the intensity of confederal competition; and, as the French case suggests, the relationship between one major confederation and one major party may be enough to set the tone. More specifically, the case studies suggest that high levels of both party control and inter-party antagonism tend to be associated with intense confederal competition, as in India. Correspondingly, confederal competition tends to be slight when inter-party antagonism is muted and party control is diffused, as in the Venezuelan case with its persisting two-party alliance in both national and confederal politics.

There are, in addition, two other factors reflected in the structural fragmentation of some party-ancillary movements which might be expected, on their own, to sharpen confederal competition. One is religion; the other is race. The first is considered below in relation to the Dutch trade union movement; the second in relation to the South African.

[19] Valente, *The Political, Economic, and Labor Climate in Venezuela*, p. 189.
[20] Ibid. 190.
[21] Ibid.

Religion: The Netherlands

Trade union confederations with an express religious or 'confessional' identification have usually been associated with the Christian religion (one exception, before Lebanon became the cockpit of the Middle East, was the Muslim confederation which operated there alongside a Christian counterpart; another is the Islamic National Labour Federation in Pakistan). In particular they have been associated with Roman Catholicism and with Catholic-linked political parties. Specifically Catholic confederations are still found in France, Belgium, Switzerland, Luxembourg, and, outside the party-ancillary category (see Chapter 13), in West Germany and in Canada.

The Canadian case, perhaps more appropriately classified in terms of a racial identification, is touched on in the next section. The West German and the French confederations are both of marginal industrial and political significance. Belgium, as we have seen, boasts the one confessional confederation which is larger than any of its competitors. Its counterpart in Luxembourg is the second largest of three. Switzerland, for its part, has not only a Catholic confederation, but a smaller Protestant one as well. They are, however, numerically outstripped, individually and in combination, by two of their three secular competitors. But there is a different story to be told about the two similar confederations which existed in the Netherlands until quite recently.

The basic structural pattern of the Dutch trade union movement was established by the time the First World War began. The earliest confederation, the Netherlands Federation of Trade Unions (NVV), associated with the Social Democratic party, was in the field by then. Alongside it were the Christian National Trade Union Federation (a Protestant confederation) and the Bureau for Roman Catholic Trade Unions. Together, these confessional confederations could claim a membership amounting to no more than half the NVV's size, but they were each linked to a party of parliamentary substance.

This pattern solidified following the Second World War. Before the war ended, a co-ordinating Council of Trade Union Federations was formed by leading members of the inter-war NVV, the Protestant confederation and the Catholic confederation—the last taking until 1964 to settle on the title by which it is now best known, the Netherlands Catholic Trade Union Federation (Catholic confederation). The co-ordinating council was opposed by a Trade Union Unity Centre, in which Communists were the prime movers; but the Centre was soon a spent force, although it lingered on for a decade or so.

By the 1970s the Socialist NVV, the Catholic confederation and the Protestant confederation together covered eight out of ten organized employees. The two confessional confederations accounted for 51 per cent of this combined membership. Each confederation had intimate, if informal, political party links: the NVV with the Labour party, the Catholic confederation with the Catholic People's party, and the Protestant confederation, farthest to the right, with the Anti-Revolutionary party and the Christian Historical Union. All of these parties, in the context of the Dutch electoral system and its penchant for producing coalition governments, have been among 'the main contenders for office'.[22]

The ability of the main parties to arrange coalitions was echoed in the relations of the three confederations with each other. As far back as the First World War, it appears, the competition between them was muted, being characterized by 'mutual tolerance and co-existence', rather than 'a ruthless struggle for exclusive representation'.[23] The co-ordinating council they formed as the Second World War was ending amounted to a further step along the same path. The principle of co-ordination was, moreover, replicated right down to the local level in formal terms—although, in practice, it was most effective at the confederal and national union levels. At the same time, in an even more striking demonstration of the co-operative spirit, the three confederations agreed to take steps to minimize 'price and quality competition' between their affiliated unions by equalizing membership dues, and redrafting formal prescriptions of 'the rights and obligations' of unionists in order to 'achieve similarity of expression'.[24] They also agreed that no affiliated union would accept a membership application from an employee who had been expelled by any of their affiliates.

In 1945, too, the three confederations had joined with all major employers' organizations to form the Foundation of Labour—from which the Communist-dominated Trade Union Unity Centre was excluded. The Foundation claimed the right to be consulted by

[22] Kendall, *The Labour Movement in Europe*, p. 264.
[23] Windmuller, *Labor Relations in the Netherlands*, pp. 138–9. [24] Ibid. 140.

government on matters of social policy. To some extent its role was taken over by the tripartite Social-Economic Council which the government set up in 1950 as its official advisory body in economic and social affairs: the same employee and employer organizations were represented on it, along with government officials. The Council and the Foundation have shared, in varying ways, the negotiation of central wage arrangements and their application at the level of collective bargaining agreements. On other issues, right down to the plant level, the three confederations and their affiliated trade unions commonly—though not invariably—acted in concert.

In 1954 there was a hiatus, a hiccup as it turned out, in this co-operative meshing. A pastoral letter directed Catholic employees to avoid or withdraw from membership of NVV unions. The NVV, in turn, withdrew from the confederations' co-ordinating council. Nevertheless, informal consultations continued until 1958, and then a loose Consultative Board brought the three confederations formally together again. The board spawned counterparts at the level of the national unions affiliated to each confederation. The Catholic hierarchy withdrew the ban on membership of NVV unions in 1965. Two years later the three confederations officially published a joint 'Programme of Action'.

These close inter-confederal arrangements, once described as a 'workable alternative to organic unity',[25] were not enough for some. During the early 1970s there were extensive negotiations on the issue of forming a single, united confederation. In the upshot, the Confederation of the Netherlands Trade Union Movement (FNV), which emerged in 1976, absorbed only the Socialist NVV and the Catholic confederation, the latter's accession owing a great deal to earlier changes affecting the hierarchy of the Dutch Catholic Church. The more conservative Protestant confederation decided to stand out. Since then, however, with the affiliation to it of a number of Catholic trade unions, this confederation has ceased to be exclusively Protestant in character—thus justifying the wider implications of its formal title: the Christian National Trade Union Federation. In this it replicates developments at the party-political level, where the party it is now associated with, the Christian Democratic Appeal, is also a Catholic–Protestant amalgam resulting from a merger, ultimately formalized in 1980, of the three

[25] Ibid. 142.

old religious parties (the Catholic People's party, the Christian Historical Union, and the Anti-Revolutionary party).

The mid-1970s also produced a separate non-party trade union confederation catering exclusively for white-collar unions. But the FNV accounts for 70 per cent of the union members covered by the three current confederations. At the same time, the FNV's emergence has plainly choked the remarkably strong co-operative element characterizing Dutch confederal relationships up to the mid-1970s. Relations between it and the Christian (once Protestant) confederation have become notably more distant. On the other hand, there is no ground in this—or in any other aspect of the Dutch case—for concluding that the religious factor has significantly sharpened confederal divisions there.

Race: South Africa

Communal or racial distinctions may be reflected in trade union structures in various ways. In Belgium, for example, the Catholic confederation, with over four-fifths of its membership drawn from Flemish-speaking Flanders, has been able to maintain a highly centralized structure. The Socialist confederation, on the other hand, with a closer balance between its Flemish membership and its French-speaking Walloon membership, has moved towards more decentralized constitutional arrangements. Elsewhere, in the heady period following the Czechoslovak Spring, the official trade union confederation of a Communist state responded to Slovak agitation by temporarily restructuring its constituent unions in ethnic terms (see Chapter 10).

It is relatively rare, however, for communal or racial divisions to be directly and explicitly reflected in confederal dividing lines. There is one case outside the range of party-ancillary movements, in Canada, where a confederation catering for unions of French-speaking workers is confined to the province of Quebec; but the regional limitation, combined with a comparatively small membership, deprives it of taxonomic significance (see Chapter 13). Among party-ancillary movements, the same regional limitation applies also in Sri Lanka and in Spain. In Sri Lanka the Federal party, based on the Tamil ethnic minority, is one of a number of parties, ranging across the political spectrum, which have set up their own trade union organizations. That associated with the Federal party, however, is less a confederation than a trade union proper since it is concerned virtually solely with recruiting Tamil plantation workers, as well as being restricted to the south-central hill country. In Spain, on the other hand, there are a number of confederations which define themselves in communal terms. But, again, all are regionally restricted, most notably to the Basque country and to Catalonia. In any event, the party-ancillary classification of Spanish trade unionism, as in the Sri Lankan case, is more conclusively justified in other terms. There are two major national confederations, linked to the Communist party and the Socialist party, respectively, together with at least four smaller non-regional confederations similarly defined by party allegiance.

There is, however, one country in which the racial factor pervades and dominates both the structure and the concerns of the trade union movement. It is South Africa. At the beginning of the 1970s, Whites constituted some 22 per cent of the employed work force, Coloureds (mixed race) almost 11 per cent and Indians something over 2 per cent; the remaining 65 per cent were Blacks. In 1972, there were 178 trade unions registered under the Industrial Conciliation Act, with a total membership of more than 600,000. None of these unions admitted Blacks. Half of them, accounting for 60 per cent of the total membership, catered exclusively for Whites; a number of others catered exclusively for Coloureds (11 per cent of the total). The remainder were 'mixed unions' enrolling Whites, Asians (Indians) and Coloureds. Including those in mixed unions, Whites made up 70 per cent of the total union membership.

About the same time, at least two Black trade unions were known to exist (with a total of only 16,000 members), although there were almost certainly more. But information about Black unions was difficult for White observers to come by, partly because such unions were not eligible to register under the Industrial Conciliation Act, and were not legally recognized in any other way. Nevertheless, Black trade unionism has a rather more substantial history than this might suggest. It is, however, a history of quite extraordinary organizational instability.

White trade unionism in South Africa traces its origins back to the nineteenth century. Its Black counterpart dates from 1919 and the foundation of the Industrial and Commercial Union, which also recruited Coloureds. This union, centred originally on dock-workers at Cape Town, soon enlarged its industrial coverage, its aims, and its structural arrangements. By 1927 it was claiming 100,000 members throughout the country. It tried to affiliate to the White-dominated trade union confederation on two occasions, in 1919 and in 1927, without success. Thereafter, for a variety of reasons, it soon petered out.

In the meantime, the Industrial and Commercial Union had inspired the emergence of a number of industrially more selective Black unions, all associated with the Communist party. In 1928 they formed the first Black confederation, the Non-European Trade Union Federation (with a mere 10,000 members); it did not survive the onset of the Depression. A resurgence in Black unionism, under the guidance of a White Trotskyist, resulted in the formation of a short-lived Joint Committee of African Trade Unions in 1938. A Council of Non-European Trade Unions, formed four years later, was longer-lived; and, during the war years at least, was 'the most powerful African trade union grouping ever to have existed in South Africa' up to the 1980s.[26] But in 1946 a breakaway, alleging undue Communist influence, produced a Council of African Trade Unions.

Earlier, White trade unionism, after one or two false starts, had by 1931 thrown up an effective and stable central confederation in the South African Trades and Labour Council. This, from its beginning, invited the affiliation of Black, as well as Coloured, mixed, and White, trade unions—although, by the mid-1940s, only seven Black unions had accepted the invitation. In 1947 some of the opponents of this open-handed policy broke away to form the Co-ordinating Council of South African Trade Unions; this denied affiliation to any union in which Blacks, Coloureds, or Asians held voting power. In 1951 a second breakaway produced the South African Federation of Trade Unions, which debarred Black unions but accepted mixed unions of Whites and Coloureds.

Three years later, in 1954, the South African Trades and Labour Council itself moved to reject Black trade unionism. It reconstituted itself as the South African Trade Union Council—later renamed and now much better known as the Trade Union Council of South Africa (TUCSA)—and confined its affiliations to White, Coloured, and mixed unions. In 1957 there was a momentary reconciliation between TUCSA and the two breakaways of 1947 and 1951, when

[26] Lodge, *Black Politics in South Africa since 1945*, p. 18.

the three joined forces in the South African Confederation of Labour, but TUCSA withdrew from the confederation within a few months.

When the old trades and labour council was reconstituted into what eventually became TUCSA, a group of its former affiliates linked up with the Black Council of Non-European Trade Unions and created a new confederation, the South African Congress of Trade Unions (SACTU) open to unions of all races. SACTU was effectively the creation of a political party, the Congress Alliance, a cluster of political organizations which included the African National Congress. SACTU was subsequently represented on the Alliance's national executive. The backdrop to SACTU's formation, and the reconstitution of the South African Trades and Labour Council into the more narrowly based confederation eventually known as TUCSA, was the passage of the Bantu Labour Act of 1953 and the introduction of the bill that was to become the Industrial Conciliation Act of 1956. They were measures that reinforced the exclusion of Black workers from the law applicable to other employees. In particular, they not only denied Black trade unions legal registration but debarred Black workers from admission to a legally registered trade union. Registration was open to Coloured and Indian unions, as well as those of Whites, but no longer to mixed unions. The Industrial Conciliation Act also formally introduced the principle (finally abandoned only in 1987) of reserving specified types of work for specified racial groups.

By 1961 SACTU was at its peak. Although formed as a multi-racial confederation, it was 'largely a multiracial head with an African body'.[27] While its claimed affiliated membership of 53,000 extended to all racial groupings (including 498 Whites), more than 70 per cent were Black; and its organizing activities were primarily among Blacks. By 1961 also, it faced a non-Communist Black confederation formed two years earlier by former TUCSA affiliates which had declined to join SACTU. This was the Federation of Free African Trade Unions (African Federation).

Subsequently, both the African Federation and SACTU were hit by different events which spelt the end for each of them. The African Federation's membership was drained by TUCSA's decision in 1962 to accept the affiliation of Black unions; it was wound up

[27] Feit, *Workers without Weapons*, p. 33.

four years later. SACTU was also, if less severely, affected in the same way by TUCSA's decision. But it was mortally wounded when the government strengthened the Suppression of Communism Act in 1965 and then used it to arrest more than 160 SACTU officials. By the late sixties SACTU had effectively disintegrated. By then, too, partly as an outcome of government pressure, the issue of Black affiliation had been the subject of mounting controversy within TUCSA; and a number of unions, mainly White, had disaffiliated. In 1969 TUCSA reverted to its tribal origins. It decided to exclude from affiliation trade unions with Black members.

By the start of the 1970s, there was thus no confederation which accepted unions enrolling Black employees. The two existing confederations were the all-White South African Confederation of Labour, which supported government policy on Black trade unionism and the principle of job reservation; and TUCSA which did not. The bulk of TUCSA's slightly larger affiliated membership was Coloured or Asian.

A resurgence of Black trade union organization followed a sudden and unprecedented wave of strikes during the first three months of 1973. These strikes, illegal like all Black strikes, were distinguished not only by the absence of violence, but also by the fact that many of them succeeded in winning wage rises. TUCSA once again decided to admit Blacks, but only if they were organized in so-called 'parallel' unions sponsored and effectively administered by affiliated non-Black unions. Black union organization received further stimulation in 1979 when the government legislated to give Black and non-racial trade unions, for the first time, the same rights of registration and collective bargaining as their White and Coloured counterparts. It is noteworthy that trade unions were singled out in this respect. Such statutory encouragement 'does not hold for any other organisations in the Black community'.[28]

By the end of 1983 the total membership of registered unions amounted to about 1.4 million, as compared with 600,000 in 1972 when the count was limited to Whites, Coloureds, and Asians. There were also, alongside the South African Confederation of Labour and TUCSA, two wholly or predominantly Black confederations. One, the Council of Unions of South Africa (Council of

[28] Lever, 'Trade Unions as a Social Force in South Africa', p. 39.

Unions), formed in 1980, was the Black counterpart of the all-White South African Confederation of Labour in that it enrolled only Black unions. The larger of the two, the Federation of South African Trade Unions, founded in 1979, was multi-racial in intent, although mainly Black in fact. In December 1985 this federation reshaped itself into COSATU, the Congress of South African Trade Unions, enlarged its affiliated membership to a claimed 450,000, and openly associated itself with the banned African National Congress and the legal United Democratic Front. COSATU's emergence evoked the creation of yet another Black (though formally multi-racial) confederation, when the Zulu-based Inkatha movement announced the foundation of the United Workers' Union of South Africa early in 1986.

The forms of confederal competition in the South African trade union movement differ from those common to other party-ancillary movements. Two factors largely account for this difference. One, of course, is the factor which overshadows all others in South African society: race. The other, much less obtrusive, stems from that part of the South African heritage which may be described as Anglo-Celtic: it has to do with the terms in which trade unions otherwise define their field of recruitment.

The racial factor means that in virtually all industries, and in a rapidly expanding range of occupations, there is room for at least two and up to four (Black, White, Coloured, Asian) trade unions defined in racial terms. But the terms of the definition eliminate inter-union competition, so far as recruitment is concerned. For it follows that when unions plying their trade in the same industry or occupation define their recruitment zones in terms of different racial groups, they necessarily exclude any competition for members. In South Africa, indeed, a number of Black trade unions have owed a great deal to the help given them by parallel non-Black unions. But, of course, racially defined unions may still compete about other things than members. In South Africa the competition focuses on differentials concerning pay and conditions. In other countries, these differentials tend to relate to differences in skills; in South Africa skills tend to be equated with race. But, more than that, there has also been the issue of reserving specific types of work for a specified race.

The Anglo-Celtic heritage of South African trade unionism reveals itself, above all, in the extent to which occupational categories provide the basis of trade union organization. The heritage extends, moreover, to the tendency of employers to recognize in relation to the industry or the occupation, as appropriate, only one trade union—which may mean, in the South African case, one for each racial grouping. Again, this entails the elimination of inter-union, and by the same token, confederal competition for rank-and-file members in cases where union representation rights have been established. At the same time, confederations may still compete at this level when it comes to moving into unorganized occupations or industries or plants, up to the point at which a new union achieves recognition. There is, as well, another level of confederal competition which this kind of union structure entails in the case of confederations affiliating unions within the same racial category. The competition in this case is for the trade unions themselves, as affiliates.

Thus, on one side, the all-White South African Confederation of Labour and TUCSA, with its mixture of affiliates, are in competition not merely in relation to the White unions affiliated to either of them, but in relation to the quite large groupings of independent (i.e., unaffiliated) White trade unions. And the lines of their competition are drawn in essentially political terms. The South African Confederation of Labour, with its strong Afrikaner element, leans decisively towards the Nationalist party and its policies. TUCSA, more liberal in its policy stance, leans at least toward the major opposition party, the Progressive Federal party. In neither case are there formal party–trade union links. Since 1956, indeed, registered trade unions have been debarred by law from using their funds for political purposes.

On the other side, TUCSA, by virtue of its 'parallel' Black unions, is also competing with the predominantly Black COSATU and its affiliates. But TUCSA–COSATU competition extends beyond Black unionism. For COSATU's claim to be multi-racial means that TUCSA's White, Coloured, and Asian affiliates are targeted as well. And, again, the competition turns on the respective political associations of the two confederations.

The new Black unions which emerged from 1973 'avoided any political orientation and constituted themselves from the bottom up, factory by factory',[29] in striking contrast with the earlier

[29] Lodge, *Black Politics in South Africa since 1945*, p. 328.

movements in Black union organization. The Federation of South African Trade Unions initially affirmed this neutral stance, but eventually (as we have seen) moved to align itself with the African National Congress and the United Democratic Front. That switch was ultimately registered in terms well understood in the South African context when the title 'Federation' was replaced by 'Congress' with the creation of COSATU in 1985.

COSATU, of course, competes frontally with the two exclusively Black confederations (the Council of Unions and the United Workers' Union) in relation to Black unions. Again, the competition is primarily on political grounds. The Council of Unions defines itself in relation to COSATU not only in racial terms, by its Black-exclusivist stance, but also in more conventional political terms by remaining professedly independent of all political parties or groupings. For its part, the United Workers' Union brings to the competitive confederal mix both a broadly moderating political influence, with its Inkatha identification, and a potentially explosive tribal element, with its Zulu connection.

SUMMATION: DEGREES OF POLARITY

The structural fragmentation characterizing party-ancillary trade union movements may vary significantly in two respects. In the first place, there is the number and diversity of competing confederations: the range here is from movements which have two or three confederal competitors to those with a half-dozen and more (not to mention the independent trade unions, unaffiliated to any confederation, which are found in almost all cases). In the second place, there is the more interesting issue of the extent to which confederal relations are polarized: the range here is from highly polarized movements in which competition between confederations is fierce, and often violent, to those in which co-operation rather than competition tends to dominate their relations. In the latter case, at the extreme, co-operation involves an effective integration which implies the kind of organizational unity, enveloping political diversity, that is characteristic of autonomous trade union movements (see Chapter 13).

Political divisions which are compounded by a racial dimension, on the evidence, decisively favour polarization. But political divisions compounded by a specifically religious dimension do not (at least in recent times) appear to do so. Otherwise, polarization in party-ancillary trade union movements seems to be associated with the tightness with which at least one major confederation is controlled by a political party—given that co-operative *inter-party* relations may modify the polarizing implications of such control.

10

STATE-ANCILLARY MOVEMENTS

Trade union movements of the state-ancillary type outnumber all others. They are invariable among countries of the Communist world. Among countries with non-Communist regimes, they are virtually the rule in Africa and the Middle East, and are common in Asia and Latin America.

Typically, but not invariably, a state-ancillary movement consists of a single confederation to which all legal trade unions, without exception, are organically linked. The confederation is ultimately subject, in all matters, to the control of the ruling political party and/or the state. In particular, its top officials are effectively appointed by the party or state leadership, 'election' by confederal organs being a formality. Usually, too, the confederation is protected by an effective state embargo on the formation of trade unions outside the framework it provides. Industrially, the confederation and its trade unions tend to display an overriding interest in maintaining and improving production. One indication of this is that (with only one or two exceptions) they never officially initiate, or endorse, strike action.

The Soviet Union provides what may be regarded as the classic model of a state-ancillary movement. It is a model that has been copied closely in other Communist countries, although there have been some deviations of substance, both temporary and long-term. For their part, non-Communist regimes have tended to conform less closely to the letter of the Soviet model, though closely enough to the spirit.

THE SOVIET MODEL

Soviet trade unions claim to include in their membership a massive 98 per cent of the total work force. Union members are grouped into some 31 industry-based national unions which cover all occupational groupings, including managerial staff. The unions, in turn, are incorporated within a single confederation, the All-Union

Central Council of Trade Unions (AUCCTU), which was founded a few months before the revolution of October 1917. There are no recognized union organizations outside the AUCCTU. Attempts to create independent unions have ended in the gaoling, exile, or commitment to mental hospitals of the instigators.[1]

All Soviet trade unions have an elaborate and almost completely uniform structure; it mirrors the parallel organization of inter-union bodies comprising the other components of the AUCCTU. Power within the confederal structure is thoroughly centralized.

> The council splits or combines national unions and shifts groups from one union to another. It approves the trade union budget, from the local units up to the national level. It issues instructions on elections and regulations for socialist competition, sets the dates and gives instructions for working out collective contracts in all enterprises. It instructs the unions on their duty to explain to workers important current decisions of the party and the government. It constantly holds before the unions their obligations as the 'school of communism'.[2]

The principle of 'democratic centralism' is taken, as in the case of other Soviet institutions, to underpin and justify this degree of central control. The democratic element is the opportunity to influence higher-level decisions which those at lower levels of the hierarchy are presumed to possess through elective and accountability procedures. The centralist element is that decisions reached at higher levels of the hierarchy are absolutely binding on lower levels.

The structure of the AUCCTU is precisely parallel to the structure of the Soviet Communist party. Party control of the confederation and its components is entrenched in two ways—formally, by way of party directives and statutes, which are legally binding on AUCCTU organs; and informally, by way of party members in those organs who are required to follow party directives and party policies in general. Through its members, the party maintains a constant presence at all levels of the AUCCTU's structure. At the lowest level, where rank-and-file workers are entitled to vote in elections for the members of workshop committees, 'auditing commissions' are required to draw up lists of

[1] See esp. Gidwitz, 'Labor Unrest in the Soviet Union', pp. 35–7: also Pravda in Kahan and Ruble, *Industrial Labor in the USSR*, pp. 355–6; Ruble, *Soviet Trade Unions*, pp. 3–4; Schapiro and Godson, *The Soviet Worker*, pp. 130–1.

[2] Brown, *Soviet Trade Unions and Labor Relations*, p. 87.

suitable candidates. Party officials are invariably members of such commissions. All full-time union positions, whether elective or appointive, are subject to the *nomenklatura,* or list, system whereby higher bodies in the hierarchy have the authority to confirm or overrule key appointments at the level below. The result is that at least the chairmen of union bodies are usually members of the Communist party. And the presence of party members increases with the level of organization within the AUCCTU—although they have been estimated to comprise only some 12 per cent of the total trade union membership.[3]

The AUCCTU leadership is constitutionally responsible to the All-Union Congress of Trade Unions. But its position and authority do not in fact depend on the congress. Thus, at a time when the congress was constitutionally required to meet every four years in order to elect the AUCCTU's leadership, among other things, not one congress was convened for a full seventeen years, between 1932 and 1949. Nevertheless, appointments continued to be made during this period to the chairmanship and all other offices. The reality of party control is similarly reflected in the fact that from 1929 until 1982, none of the men appointed as the AUCCTU chairman had any previous experience in the trade union movement. All were party members brought in from outside; and they included Alexander Shelepin (1967–75), the only one to be simultaneously a full member of the party's Politburo, who had headed the Committee of State Security (KGB) before moving to the AUCCTU. In any case, the guiding role of the party is repeatedly acknowledged by trade union bodies. Thus the AUCCTU's 16th congress in 1977 assured the party's Central Committee 'that the trade unions of the USSR will always be reliable supporters of the Party and its active helpers'.[4]

One aspect of the party's interest in the trade unions conforms with Lenin's prescription (see Chapter 6) that they are a 'school of communism', a phrase religiously echoed in the speeches and writings of his successors. In their daily dealings with the unions, party officials 'appear to be most directly concerned with union educational and cultural activities', especially 'political education'.[5] But if, in the long term, the party sees the schooling provided by the

[3] See Lowit, 'The Working Class and Union Structures in Eastern Europe', p. 68
[4] Quoted in Schapiro and Godson, *The Soviet Worker,* p. 111.
[5] Ruble, *Soviet Trade Unions*, p. 61.

unions as moulding a 'new spiritual type of man' (to quote Leonid Brezhnev),[6] in the short term its gaze is focused decisively on the material end of promoting production. This end, too, is repeatedly acknowledged by trade union bodies. Thus, once again, the 16th congress which claimed, as a 'characteristic feature' of the trade unions, 'their ceaselessly deepening concern with the problems of the economy [and] of production and their . . . drive towards higher efficiency and quality of work'.[7] This concern and this drive are expressed, above all, in trade union involvement in the enforcement of 'labour discipline' in relation to such things as lateness for work, absenteeism, and faulty workmanship. They are expressed in a different way in the fact that official union-initiated strikes are unheard of in the Soviet Union. Strikes, however, are known to have occurred (almost, but not quite, invariably without any form of even local union support);[8] and severe treatment has usually been meted out to strike leaders.[9]

Soviet trade unions are also concerned with production in the broadest sense through the involvement of the highest level of AUCCTU leaders in meetings with central government planning bodies, the State Planning Agency (*Gosplan*) and the State Committee on Labour and Social Questions. At this level, the participation of the AUCCTU leadership in the process of formulating national plans gives it a voice in the formation of national wage policy. Otherwise, the trade unions have nothing to do with the determination of basic wages and salaries, since the national plan fixes a wages fund for each enterprise and pay scales are also fixed centrally—although local union bodies may have a hand in calculating certain incentive payments ('socialist competition premiums') allocated from a central fund.

Apart from the planning process, the trade unions have both a legislative role and an administrative role which stamps them as governmental agencies. The AUCCTU commonly drafts, or takes part in drafting, legislation on labour matters; and it is consulted

[6] Quoted in Schapiro and Godson, *The Soviet Worker*, p. 109.

[7] Quoted, ibid. 110–11.

[8] Pravda in Kahan and Ruble, *Industrial Labor in the USSR*, p. 350.

[9] See esp. Gidwitz, 'Labor Unrest in the Soviet Union', pp. 32–5: also Pravda in Kahan and Ruble, *Industrial Labor in the USSR*, pp. 348–51; Ruble, *Soviet Trade Unions*, pp. 100–3; Schapiro and Godson, *The Soviet Worker*, pp. 78–9, 123–4; Ziegler, 'Worker Participation and Worker Discontent in the Soviet Union', pp. 247–8.

by government agencies about policy decisions. More striking, however, is the fact that the AUCCTU on occasions apparently shares in the legislative process itself. Thus the Council of Ministers (the central executive body of the Soviet state) has formally joined with the AUCCTU in issuing decrees. Sometimes, moreover, major decrees have been issued jointly in the name of the AUCCTU, the Council of Ministers, and the Central Committee of the Communist Party (though they are normally listed in the reverse order).

On the side of administration, it is the individual trade unions which are primarily involved. They are charged with the responsibility of enforcing the application of labour laws in relation to specific industries and specific enterprises. In particular, apart from their concern with labour discipline, they have comprehensive powers to enforce health and safety legislation—employing for this purpose, in addition to part-time voluntary committees of elected trade union representatives, professional inspectors whose salaries are paid from state, not union, funds. They are the sole administering authority in relation to social security policies concerning pensions, disability, and pregnancy benefit payments; and non-unionists receive only half the disability and pregnancy benefits available to union members. In addition, they have the major role to play in administering health and welfare services in general, in the allocation of housing and of places in nurseries and kindergartens; and also in the provision of sports, recreational, cultural, and tourism facilities. All of this helps explain the remarkably high union density rate mentioned earlier.

The parallel with the functions of a ministry of labour and social welfare is obvious to Western eyes. It is more than a parallel, historically speaking. For in 1933 the Commissariat (Ministry) of Labour was formally abolished and all its functions were vested without qualification in the AUCCTU. This, of course, is to present the Soviet trade union movement as nothing more than an agency of government—or, more precisely, an agency of the ruling party in a one-party state. But the 'dual functioning' theory of trade union purpose under socialism, which Lenin made part of Communist orthodoxy (see Chapter 6), also postulates a trade union role involving the protection of workers' interests.

The formal picture on the issue of the trade unions and the protection of workers' interests is quite clear. Protection of their members' interests was specified as a 'right' of the unions in the

Labour Code of 1922. But subsequently the notion of protection of workers, and specifically the term 'defence' in this connection, 'dropped out of use'.[10] The formal emphasis was wholly on production. In the late 1940s and early 1950s, trade union constitutions were explicitly nominating the *state* as the protector of workers' rights. By the late 1950s, however, the formal story had changed, and the constitutional formulation was moving towards the position it eventually reached in 1963: 'Trade unions defend the interests of workers and employees'.[11]

The extent to which Soviet trade unions in practice protect the interests of their members is a matter of some dispute. On the one hand, there are the many accounts, for example, of trade union failures to remedy safety hazards because to do so would have halted or retarded production. On the other hand, there are the contrasting stories of benevolent and successful trade union action. But the critical point, for present purposes, is that not one of those who dispute that issue call into question the reality of either Communist-party control of the trade unions or their state-like functions.

COMMUNIST VARIANTS

The Soviet model has been applied throughout the Soviet bloc of Eastern Europe, by other Communist regimes in Yugoslavia and Albania, in Cuba, in China, North Korea, Afghanistan, Mongolia, and, it appears, Vietnam and Kampuchea. For the most part, such variations from the model as have occurred in these cases have not seriously disturbed the central feature of decisive party control. But there have been some exceptions: all were temporary developments, the product of domestic crisis, and were followed by effective restoration of the Soviet model.

One such development occurred in Hungary during the revolution of 1956. Workers' councils, independent of party control, were formed in many workplaces; and, as the official trade union organization tottered, a provisional organizing committee published a programme for establishing a National Federation of Free Hungarian Trade Unions which was structurally the antithesis of the Soviet model. A few days later, however, Soviet tanks rolled

[10] Brown, *Soviet Trade Unions and Labor Relations*, p. 65.
[11] Quoted ibid.

across the border. A Central Workers' Council, formed shortly after the invasion, was within a month banned and its leaders arrested.

There was a similarly spontaneous movement to form workers' councils during the parallel, but less explosive, crisis in Poland following the Poznań strike of mid-1956. The councils were regarded as an independent alternative to the official trade union organization, which was widely described as 'the second government'.[12] Władysław Gomulka, who rode back to power in the Polish United Workers' party (Communist party) on the crest of this crisis, quickly promulgated a law legitimating workers' councils. Their legal powers, however, were circumscribed and, in part, obscure. Gomulka rejected proposals to extend them in 1957; and a year later had neutralized them by effectively subjecting them to the corresponding party and trade union organs.

The Czechoslovak Spring of 1968, signified above all in the emergence of Alexander Dubček as head of the Communist party, resulted in moves towards much more wholesale changes in the Soviet model. Unusually, instead of involving separate organizational forms and external initiators, they were changes in the structure and practices of the official trade union organization itself, and they were brought about by the official leaders (many of them, however, newly in office). Oddly enough, too, most changes occurred after the Soviet invasion in August, during the period of occupation which preceded Dubček's downfall in April 1969. In these months of the interregnum, the official confederation, the Revolutionary Trade Union Movement (ROH)—primarily through its principal administrative organ, the Central Council of Trade Unions—disavowed or modified key elements of the Soviet model. The central function of the unions was redefined as representing the sectional interests of workers in relation to management; the dependence of wage settlements on the terms of central economic plans was rejected; an ultimate right to strike was asserted; decentralization of control within the ROH itself was enhanced by shifting authority to the individual unions and by restructuring the former twelve constituent unions into fifty-seven (and, in doing so, the Slovak minority's claims were recognized by making all but one of the new unions ethnically exclusive to either Czechs or Slovaks).

[12] Quoted in Sturmthal, *Workers' Councils*, p. 125.

Moreover, there was strong if unfulfilled pressure, acknowledged by the predominantly Communist party leadership of the ROH and its constituent bodies, for the trade unions to declare their independence of the Communist party: 'Trade unions without communists' was a well-known slogan.[13] But all these developments were soon reversed once Dubček fell and the so-called 'normalization' process began.

Even more remarkable was the Polish deviation of 1980–1 when, for thirteen months, the shell of an official trade union confederation faced a legally registered competing organization which was effectively independent of the ruling Polish United Workers' party. This, the story of Solidarity, is told below in Chapter 11. Similar attempts to form independent trade unions, if on a more localized basis, have been reported from Romania (1979) and from China (1981).[14]

Otherwise, however, lasting Communist variations on the Soviet model have been a matter of minor modification within the Soviet bloc—with, it seems, a possible exception in the case of Hungary.[15] But more certainly, clearly substantial Communist variations on the model have occurred outside the Soviet bloc. The key cases are China and Yugoslavia.

China

The official trade union movement in the People's Republic of China has undergone cataclysmic, if temporary, transformation in crisis situations. At the time of the Great Leap Forward (1958–60), the All-China Federation of Trade Unions (ACFTU) was an early casualty of factional warfare within the Communist party. Trade union leaders were purged, some union bodies became little more than paper organizations and the ACFTU, along with the trade unions in general, was greatly weakened. The early sixties, despite a less hostile climate, saw no great improvement in their position. They came under attack again with the onset of the Cultural Revolution in 1966. This time, it was the final catastrophe: the ACFTU and all its constituent trade unions were formally

[13] Quoted in Windmuller, 'Czechoslovakia and the Communist Union Model', p. 41.

[14] See Pravda and Ruble, *Trade Unions in Communist States*, pp. 120–1; Wilczynski, *Comparative Industrial Relations*, p. 81.

[15] See esp. Pravda and Ruble, *Trade Unions in Communist States*, pp. 13–14.

abolished. But then, with the petering-out of the Cultural Revolution, the ACFTU was 'allowed to reopen its offices' in 1971,[16] although some years passed before it became fully operational. In the same mode, its constituent trade unions were haltingly re-established. Not until 1978, after Mao's death and the arrest of the Gang of Four, did the ACFTU convene the first national trade union congress to be held in twenty-one years.

Apart from its episodic existence and erratic organizational scope, the ACFTU diverges from its Soviet counterpart in two other particularly notable respects. In the first place, it appears to be much less highly centralized. Its formal structure displays the familiar pattern of seventeen national trade unions based on industry divisions; but the operational emphasis has come to be placed on regional and local inter-union bodies, and on organization at the enterprise or plant level. Thus, the recovery from the devastation of the Cultural Revolution began with, and depended upon, local and provincial—not industrially-focused—trade union initiatives.[17] And since then the trade union branch in the plant or enterprise has been effectively merged with a non-union workers' congress (meeting normally twice a year) in the sense that the union formally acts on behalf of the congress in between meetings.

The second notable divergence from the Soviet model is reflected in the fact that the ACFTU's membership comprises a far smaller proportion of the total non-agricultural work force: about 60 per cent. This remarkably low figure (by Communist standards) is largely, it appears, a consequence of two factors. One is that the coverage of the unions is not comprehensive: more than one-fifth of non-agricultural workers are employed in enterprises that have no union organization.[18] The other factor is that union membership, while open to all levels including management, has been confined to the category of permanent or 'regular' employees. As a result, a sizeable and apparently expanding minority of workers employed as apprentices, on fixed contracts or on a temporary or casual basis has been debarred from membership.[19] This restriction, unfamiliar

[16] Henley and Chen, 'A Note on the Appearance, Disappearance and Re-appearance of Dual Functioning Trade Unions in the People's Republic of China', p. 89.

[17] See Lee Lai To, *Trade Unions in China*, pp. 131, 160–1, 168–9.

[18] See Pravda and Ruble, *Trade Unions in Communist States*, p. 229.

[19] See Hoffman in Blum, *International Handbook of Industrial Relations*, p. 139; Walder, 'The Remaking of the Chinese Working Class', pp. 37–8.

to non-Asian Communist trade union movements, also characterizes the non-Communist movement of Japan (see Chapter 11). As in Japan, too, membership of a trade union in China carries with it specific privileges: in the Chinese case it involves welfare provisions, sporting and cultural facilities, and housing opportunities.[20] (There is the whiff of similar privilege in a reference to a third Asian—and Communist—case: workers who failed in their applications to join a union linked to the Association of Kampuchean Trade Unions were consoled with the injunction 'to work harder and improve their qualifications'.)[21] It was this distinction, between regular workers and the rest, which Mao exploited during the Cultural Revolution in the attack on his factional opponents who were strong in the trade unions. The significance of the workers' congresses, established during that time, is that they included *all* employees. On the other hand, there appear to be changes afoot which, if carried through, will diminish the importance of permanent employment and, perhaps, may eventually bring the membership requirements of Chinese trade unionism closer to those of the Soviet model.[22]

Nevertheless, whatever the future holds, the essential similarities between the two remain. One is the emphasis on the promotion of production as the primary task of the trade unions. The other, and most crucial, is the domination of the Communist party expressed both in formal constitutional provisions and, above all, in the presence of party members in all key positions.

Yugoslavia

The Confederation of Trade Unions of Yugoslavia (CTY) is structurally much like its counterparts in the Soviet Union and Eastern Europe. Like them, it monopolizes trade union organization, recruits managerial staff, and includes more than 95 per cent of the work force in the membership of its constituent unions. But, unlike them, the CTY operates in the context of a society in which the ruling Communist party, since the early 1950s, has accepted a degree of decentralization that represents a radical

[20] See White, *Careers in Shanghai*, pp. 62, 109, 116, 181–2.

[21] Vickery, *Kampuchea*, p. 118.

[22] See Warner, 'Industrial Relations in the Chinese Factory', pp. 229–30; and Pravda and Ruble, *Trade Unions in Communist States*, p. 229.

departure from the strict centralism of the Soviet system. The potent communal, linguistic, and religious divisions within Yugoslavia have played a major part in bringing this about. They are reflected directly in the federalized constitutional arrangements which, since the early 1970s at least, represent a genuine diffusion of state power among the six republics and two 'autonomous regions'.

Decentralization has even impinged on the Communist party itself. That is symbolized in the party's 1952 decision to change its name to the League of Communists in Yugoslavia. And beyond the symbolism, while the party holds still to democratic centralism as an internal imperative, and can still on occasion condemn 'localism' and 'particularism' within its own structure, it has been decisively federalized as the result of a factional struggle that, from the late 1960s, shifted the power to appoint high and middle-rank party officials from its central to its republican (regional) organs. The shift has also affected the way policy is decided. A similar federal emphasis is evident in the case of the CTY—the 'middle' level being, as in the party and the government, the 'most influential'.[23]

It is, however, the system of industrial 'self-management' (as the regime prefers it to be known) which is by far the most renowned and most dramatic aspect of Yugoslav decentralization. This system involves the exposure of industrial enterprises to market forces; and, by the same token, means that Yugoslav enterprises, unlike their Soviet cousins (before, presumably, *perestroika*), are not bound to the wheels of a central economic plan—which is not, however, to say that there is no central planning at all. But what the 'self-management' system involves, above all, is the existence of workers' councils in all 'socially' owned enterprises. In many, there are separate workers' councils for notionally distinct units known officially as 'basic organizations of associated labour', which co-operate with each other in the productive process of a particular enterprise on the basis of a 'self-management agreement' among themselves.[24] Workers' councils are elected (in a Soviet-style ballot in which the number of candidates exactly matches the positions to be filled, voters having the choice of approving or disapproving) by all workers in the enterprise or basic organization, including managerial staff. On the other hand, the formal powers that

[23] Zukin, 'The Representation of Working-class Interests in Socialist Society: Yugoslav Labor Unions', p. 290.

[24] Lydall, *Yugoslav Socialism*, p. 93.

Yugoslav workers' councils have acquired, since their inception in the early 1950s, are far wider than those conferred on corresponding bodies that have surfaced from time to time in other Communist states. They decide production and investment policy, wages and prices; they approve accounts, hirings and firings, and supervise disciplinary matters in general; they confirm the distribution of certain benefits (when available), such as housing; and they have the final approval in the case of the appointment (for a four-year term) of directors or, where appropriate, members of managing boards.

The principle of self-management, expressed in the Yugoslav workers' councils, inevitably raises not only the question of the role of the party, but even more sharply the role of trade unions. Formally speaking, the trade unions since the mid-1970s have acquired a 'protective' or representational role so far as workers' relations with management are concerned. In earlier days, this was disowned: 'We show our concern for the workers' interests', declared the CTY president in 1959, 'not so much by representing them as by educating them.'[25] Up to the mid-1970s, the unions' formal role at the enterprise level was largely restricted to a limited welfare function (helping with individual workers' financial problems), and the organizing of ceremonial, cultural, and educational activities. In addition, they were entrusted with the task of organizing workers' council elections and, in particular, nominated the candidates—although, a qualification of questionable significance, a non-official group might offer an alternative list in certain circumstances.

A weightier formal role was conferred on the trade unions in the mid-1970s. They were to be regarded, it was affirmed, as 'a "partner" rather than a subordinate of the self-management organs'.[26] The result is that the trade union within the enterprise now has a major voice in the selection of the director, and is entitled to propose his dismissal; as before, it nominates the candidates for election to the workers' council; it shares with the workers' councils and their executive bodies all planning functions; and it is one of the signatories to all agreements regulating business relationships between 'basic organizations of associated labour'

[25] Quoted in Singleton, *Twentieth-century Yugoslavia*, p. 137.

[26] Zukin, 'The Representation of Working-class Interests in Socialist Society', pp. 300–1.

within the same enterprise. In addition, the unions' functions have been extended to the legal representation of workers on personal issues, such as dismissal, and to the enforcing of occupational health and safety laws. Outside the enterprise, too, the trade unions are allotted positions in the structure of assemblies, reaching from the communal to the federal level, which is the main feature of the legislative side of Yugoslav government.

The formal changes of 1974–6 were designed to give the trade unions a more impressive image than they had previously enjoyed, eclipsed as they had been—in the eyes of workers—not merely by the party, but more particularly by the workers' councils. As a result of these changes, they were 'supposed to be autonomous of party and state control'.[27] At the same time, their wider powers made them more useful to the party and, specifically, enhanced their importance as part of the mechanism by which the party controls the workers' councils.

The reality of the workers' councils, in any case, is that in general their proceedings are dominated by managerial and technical staff members, despite the numerical preponderance of manual workers on them; and managements usually secure the decisions they want on matters of concern to them. But behind this specifically managerial guidance of workers' councils there are what Harold Lydall describes as the two major 'levers' by which the party exerts authority over the enterprise, its management, and its workers' council.[28] One lever is the party's control over the banks. The other is its control over 'the political *aktiv*' within the enterprise, the *aktiv* being a small informal grouping consisting of party members occupying major posts in the workers' council, in management, and in the party and trade union organizations within the enterprise. The trade union secretary is usually a leading member of the *aktiv*. Like the chief director of an enterprise, he is invariably a party member, and his appointment is screened by party organs outside the enterprise, usually at the city level. But his likely standing within the *aktiv*, in relation to the party leadership, is tellingly implied in an observation based on a study of two factories. The local organs of the party usually sent their secretary or another representative to the meetings of all other bodies, including the trade unions—and frequently 'passed judgement upon the other bodies, but the latter

[27] Ibid. 289. [28] See Lydall, *Yugoslav Socialism*, pp. 115–18.

did not comment upon the work of the [party's] factory organization'.[29] The superior status of the party is evident in this.

In general, it has been claimed, the wider functions of the unions have not entailed greater independence: 'Underneath the new rules and regulations, the labor unions have really been brought closer to the party'.[30] More bluntly, it has been put that the trade unions are 'a subservient agent of the Party, which appoints their officers and determines their policies'.[31] But the point to be made is that the implications of such a relationship (for trade union behaviour) are not quite as clear-cut in Yugoslavia as they tend to be in the other Communist states which conventionally place far greater weight on the centralization principle.

One distinctive feature of the Yugoslav system, so far as Communist regimes are concerned, is what appears to be (the data available in English-language sources are sketchy) the more or less continuing incidence of strikes since the late 1950s. This pattern stands in stark contrast to the rare and strictly sporadic outbursts mentioned earlier in relation to Poland, the Soviet Union, and others. The actions involved, if mayhem in Soviet terms, tend to be quite limited by Western standards. During 1980, for example, an average of 'about 50' strikers took part in just 245 stoppages which lasted, on average, 'only a few hours'.[32] On the other hand, the scale of strike action was sufficiently impressive, in March 1987, to persuade the government to amend a law imposing wage cuts; and the earliest known post-war strike, in January 1958, involved as many as 4,000 employees in the coal industry.

There is no record of a trade union officially initiating a strike, though one local union once seems to have at least offered to organize a four-hour stoppage in a car factory.[33] Initiation apart, however, there is some evidence of union *support* for strikes. Such support was implied in the case of the path-breaking coal strike of January 1958, in that the participants included all leading officials of the enterprise's union organization. But even more arresting is the evidence of a survey concerning 512 strikes, most of which occurred during 1966–9.[34] The enterprise unions involved either

[29] Kolaja, *Workers Councils*, p. 38.

[30] Zukin, 'The Representation of Working-class Interests in Socialist Society,' p. 305.

[31] Lydall, *Yugoslav Socialism*, p. 129.

[32] Ibid. 122 n.

[33] See Carter, *Democratic Reform in Yugoslavia*, pp. 206–7.

[34] See Jovanov, 'Strikes and Self-Management', p. 371.

opposed both the strike action and the strikers' demands or (amounting to the same thing) declined to take a position in 44 per cent of the strikes; and they opposed the strike, but supported the strikers' demands, in another 45 per cent. But the remarkable feature of the survey was that in the case of the remaining 32 strikes (11 per cent), the trade union openly supported not merely the strikers' demands, but also their use of the strike weapon. Moreover, this highly unusual reaction at the local level, by official union bodies in a Communist state, finds an echo in a comparatively tolerant response to strikes at the upper levels of the party and the state in Yugoslavia.[35]

NON-COMMUNIST VARIANTS

State-ancillary trade union movements in non-Communist countries all share one characteristic that distinguishes them from their Communist counterparts. They do not enrol managerial staff in their membership. Otherwise, they usually follow the Soviet model in its essentials: monolithic structure, production-oriented functions, and party–state control of key appointments. There is, however, a tendency towards less elaborate organizational arrangements, a narrower range of general union functions, and simpler mechanisms of party–state control (though not often quite as brutally simple as the device, used by military regimes in Argentina and Brazil, of installing 'military overseers' in hundreds of union bodies).[36] In particular, there are more extreme deviations from the Soviet model than in the case of Communist trade union movements, in relation to structure and to control mechanisms. The range is illustrated by the cases of Tanzania and Mexico.

Tanzania

Both the political system and the trade union movement of Tanzania display key characteristics of the Soviet model, although otherwise there are great variations arising especially from differing social and economic circumstances. There is also a constitutional variation which needs to be taken into account.

[35] See Carter, *Democratic Reform in Yugoslavia*, pp. 205–7.

[36] Frank, *Capitalism and Underdevelopment in Latin America*, p. 215; see also Alba, *Politics and the Labor Movement in Latin America*, pp. 233, 239.

The United Republic of Tanzania was formed in 1964 by a merger of the newly independent states of Tanganyika (1961) and Zanzibar (1963), the latter consisting of the islands of Zanzibar and Pemba. Constitutionally speaking, it was a curious union. For Zanzibar, with less than 3 per cent of the total Tanzanian population, retained complete control of its own internal affairs, while guaranteed at least the national vice-presidency and almost a third of the seats in the national legislature. The story of Tanzanian trade unionism, as of most other things in Tanzania, is thus primarily a matter of the mainland, Tanganyika—which is accordingly the focus of this section.

The Tanzanian political system is a frankly one-party system; and was constitutionally enshrined as such in 1965. The Tanganyika African National Union (TANU), the ruling party from independence, became the Revolutionary Party of Tanzania (CCM–Chama Cha Mapinduzi) in 1977 when TANU formally amalgamated with the Afro-Shirazi party, its counterpart on Zanzibar. All candidates for election to the National Assembly must be approved by the CCM, which usually endorses two candidates for each single-member seat. Almost a third of the assembly's members are appointed either by the assembly itself, the Zanzibar legislature or the national president.

The trade union movement, equally monolithic in character, is intimately associated with the party. Up to 1964, the two were organizationally independent. Despite a pronounced co-operative element that had developed between TANU and the original trade union confederation, the Tanganyika Federation of Labour (Federation of Labour), there was also a great deal of friction in the relationship once the party shouldered governmental responsibilities from the late 1950s. Attempts to improve relations led in 1960 to the appointment of the Federation of Labour's president, Rashidi Kawawa, as a government minister; and later to the formal allocation to it of two seats on the party's executive (though the confederation consistently rejected suggestions of formal affiliation with the party, either for itself or its member-unions). In 1962, Kawawa, now Prime Minister, took the process a stage further. He made his successor (as Federation of Labour president) the Minister of Health and Labour, his chief trade union opponent the High Commissioner in London, and appointed other trade union leaders to administrative and political posts. He also passed legislation which effectively outlawed strikes and made the legal recognition of trade unions dependent on their being affiliated to the Federation of Labour.

In the upshot, the unions abandoned neither their opposition to key government policies nor their use of the strike weapon—despite the imposition of penalties under the new legislation, including the arrest and 'rustication' of some trade union leaders. Matters came to a head in January 1964 when the army mutinied in support of a pay rise and the Africanization of the officer corps. These claims were 'interpreted by the [TANU] government as signs of union influence among the mutineers'.[37] The mutiny, put down with the help of British troops, was followed by the arrest of the most prominent trade union leaders as well as many union members. It was also followed by the forced dissolution of the Federation of Labour and its affiliated trade unions.

In their place the government set up a single National Union of Tanganyika Workers (National Union) which was reorganized in 1978 to cover Zanzibar, and renamed the Union of Tanzania Workers (JUWATA). JUWATA, like its predecessor, the National Union, is highly centralized: its constituents are officially described not as unions, but as 'sections'. Apart from a so-called 'Workers' Department' of the ruling party (CCM) on Zanzibar, trade unionism can legally exist only within JUWATA's framework. Private employers are required to apply the check-off system and collect membership dues on its behalf, and the distribution of funds between the JUWATA sections and headquarters is determined by the government. The appointment of both its secretary-general and his deputy is vested in the Tanzanian president. The first secretary-general of the National Union whom Julius Nyerere appointed in 1964 was simultaneously Minister of Labour (not, it should be said, that this is a peculiarly African arrangement: for example, the Romanian Minister of Labour was simultaneously president of his country's trade union confederation from 1977 to 1981).

JUWATA's formal relationship with the ruling party is technically different from that of its immediate predecessor. Whereas the National Union was formally affiliated to TANU, JUWATA is constitutionally an outgrowth of the CCM, being simply 'one of [the party's] mass organizations'.[38] Its formal objectives are heavily

[37] Tumbo in Tumbo *et al.*, *Labour in Tanzania*, p. 18.

[38] Mihyo in Damachi, Seibel, Trachtman, *Industrial Relations in Africa*, p. 252.

oriented towards the promotion of production, and the behaviour of its local officials reflects this emphasis.[39] Tanzanian trade union organizations, as a result, 'are now regarded as an arm of the government and of the party'.[40] Nyerere's belief in the necessity of this kind of trade union movement is plain in the way he hailed the formation of the National Union: 'Thus, the industrial side of the labour movement was ... prepared for its role in a socialist economy even before much other progress had been made in that direction.'[41]

Mexico

The Mexican political system, commonly described as authoritarian and identified with one-party systems by its students, has the appearance of a multi-party system. The lower house of the Mexican congress, in 1979–82, contained representatives of the governing party, the Institutional Revolutionary party (PRI), and six others—two on the right of the PRI and four (including the Mexican Communist party) on the left. These parties had contested elections for congress, for state governor and for national president, in which voters had been free to choose between their candidates. Moreover, the successful candidates, whether members of congress, governors, or the president, were all restricted to one term in office—a constitutional restriction which, since 1940, has meant substantial changes in the top personnel of national and state governments at least every six years.

The Mexican trade union movement, too, does not have the appearance of the state-ancillary type characteristic of authoritarian political systems. Instead of the standard monolithic structure, there is an array of confederal organizations, the more significant of which are the Mexican Workers' Confederation (CTM), the Federation of Civil Service Unions, the Regional Confederation of Mexican Workers, the Revolutionary Confederation of Workers and Farmers, the General Confederation of Workers, the Authentic Labour Front, the Federation of Independent Unions and the General Union of Workers and Farmers. Breakaway unions, a frequent phenomenon in recent times, may be

[39] See Samoff, *Tanzania*, pp. 189–90.
[40] Tumbo in Tumbo *et al.*, *Labour in Tanzania*, p. 19.
[41] Nyerere, *Freedom and Development*, p. 274.

legally recognized and their founders are not persecuted as a matter of course. Inter-union competition for members is common since jurisdictional monopolies are not enforceable by law. Within unions, rather than imposing some version of 'democratic centralism', the Mexican department of labour has intervened on behalf of dissident rank-and-file groupings; and in the automobile industry, for example, there are 'several unions . . . characterized by lively and meaningful internal democracy'.[42] The right to strike, although technically circumscribed, is legally available. More to the point, strikes (both legal and illegal) are common, and so is collective bargaining leading to agreements with private employers. In general, that is to say, 'the Mexican state appears to allow considerable autonomy to many unions'.[43]

The contrast with the Soviet model or the Tanzanian case is striking and genuine; but there is another side to it. It is reflected in the phrase 'official party', a common way of referring to the Institutional Revolutionary party. The PRI is the dominant party in terms of both membership and electoral support. Since 1940 it has always won a majority in congressional elections. In the 1979–82 congress it held 296 of the 400 seats available. Its nearest rival, the right-wing National Action party, held 43, leaving 61 to be divided between the five other parties. Moreover, while *all* of the PRI's seats and just four of the National Action party's had been won on the basis of the popular vote in single-member constituencies, the balance (100) had been allocated on the basis of a supplementary proportional representation scheme which had been introduced, in 1977, precisely in order to increase non-PRI representation. As in congressional elections, so in elections for the key positions of state governor and national president: the opposition parties never win. There is thus no question of the PRI's dominance. But its relations with the opposition parties are complex, as indicated by the introduction of the proportional representation component in congress. For more than forty years the PRI's leadership has skilfully varied its handling of opposing political groupings, having 'defused them in times of danger, regenerated them when opposition was embarrassing by its absence . . . [and] used both cooptation and coercion'.[44]

[42] Roxborough, *Unions and Politics in Mexico*, p. 143.
[43] Ibid. 26.
[44] Kenneth M. Coleman, *Diffuse Support in Mexico*, p. 14.

The realities of the trade union movement underline the PRI's dominance. There is a great deal of uncertainty about union memberships and union density rates in Mexico; but there is no question that the CTM is by far the biggest confederation, with a probable membership of about two million, and covers something in the order of two-thirds of all union members. The Federation of Civil Service Unions, which includes teachers and is legally debarred from affiliating with any wider confederation, has a membership of more than 400,000. The other confederations dwindle down from an estimated actual membership of less than 150,000 to a few thousand. The CTM and the Federation of Civil Service Unions are each the major component of one of the three formal 'sectors' into which the membership of the PRI is divided—the agrarian sector, covering associations of those benefiting from government land reforms; the popular sector, aimed at the middle class, in which the Federation of Civil Service Unions is predominant; and the labour sector, in which the CTM is the doyen. In both cases formal affiliation with the party is involved. Most of the other trade union confederations are also affiliated to the PRI. With the CTM and the Federation of Civil Service Unions, they comprise what are commonly called the 'official trade unions'. Their leaders are often appointed by way of congressional (and that means, essentially, presidential) nomination.

The PRI has made a number of attempts to unify and centralize trade union organization in the past. One outcome is a kind of super-confederation, the Labour Congress, which encloses all confederations and major independent unions affiliated to the PRI through its labour sector. Alongside the CTM in the Labour Congress are three of the other confederations mentioned earlier: the Regional Confederation of Mexican Workers, the Revolutionary Confederation of Workers and Farmers, and the General Confederation of Workers. The Congress, however, is a loose body which has not obviated rivalry between the CTM and the three smaller confederations.

In addition, there is the source of inter-union competition provided by union organizations which are not affiliated to the PRI at all, and which nevertheless operate openly and legally. They are especially a development of the 1970s, and originated usually as breakaways from the CTM. Many of them came together to form the three other confederations mentioned earlier, the Authentic

Labour Front, the Federation of Independent Unions, and the General Union of Workers and Farmers, all of which remain outside the Labour Congress and the PRI. This segment of the trade union movement, it must be said, covers only a very small minority of the organized work force. Nevertheless, its significance is that it reflects a rank-and-file volatility which the Mexican authorities have experienced difficulty in containing despite sometimes draconic measures of repression. Thus: 'One area where political control is not total is precisely in what is regarded as the principal pillar of the regime, organized labour.'[45] On the other hand, partly because they fear government repression, the breakaway confederations and their unions are careful to remain 'formally apolitical', and concentrate on collective bargaining activities[46]—but in doing so, they are often involved in attacking less militant and allegedly corrupt CTM union officials.[47]

It is the CTM, however, which unquestionably remains the dominant trade union body industrially and, above all, politically. It has 'privileged access to the highest levels of political decision-making', including the president himself.[48] Its leading members find their way into a wide range of senior government positions, and its leadership is said to exercise considerable influence in the selection of the PRI's new presidential candidate every six years. In short, its intimate relationship with the PRI is such as to ensure that 'its presence is felt in virtually every federal ministry, every state and municipal government office, and most industrial plants'.[49] Nevertheless, the CTM differs in one very distinctive way from confederations more typical of state-ancillary movements: it has used the threat of national strikes as a successful bargaining counter in dealings with major employers.[50] It is not clear, but seems highly likely, that there have also been strikes initiated, or at least endorsed, by the CTM. On the other hand, it is certainly clear that the CTM has generally been inclined to favour a less militant, more production-oriented approach; and that most strike action in

[45] Roxborough, *Unions and Politics in Mexico*, p. 165.

[46] Davis and Coleman, 'Labor and the State: Union Incorporation and Working-class Politicization in Latin America', p. 400.

[47] See Schlagheck, *The Political, Economic, and Labor Climate in Mexico*, pp. 121–3.

[48] Zapata in Blum, *International Handbook of Industrial Relations*, p. 359.

[49] Schlagheck, *The Political, Economic, and Labor Climate in Mexico*, p. 115.

[50] See ibid. 138–9.

Mexico is either unofficial or the work of union leaderships outside the CTM. The CTM has commonly opposed such strikes—even, on occasion, to the extent of urging that the army be brought in.[51]

SUMMATION: DEGREES OF UNITY

It is unity which characterizes state-ancillary trade union movements, in contrast with the fragmentation characteristic of their party-ancillary counterparts. In the vast majority of cases, that unity is reflected in a confederal structure which is all-inclusive and tends to be highly centralized. Where it is not, as in Mexico, the reality of a structurally fragmented trade union movement matches the reality of an ostensibly multi-party political system. Both, while not entirely façades (since they do admit an element of genuine if constrained competition), are subject to the ultimate imperatives of the party-state. In the case of the trade unions, those imperatives are expressed primarily through the dominant confederation, interlocked as it is with the party and the state.

The crucial point, however, is not so much the structure as such, but rather the tightness of the party-state control exercised on and through it. The most convincing yardstick of loose as against tight control is the ability of the official trade union movement, especially at lower levels of the structure, to adopt a less than rigid position on the issue of productivity and strikes. Thus, in the case of Communist political systems, there is on this yardstick a clear distinction between the Yugoslavian movement and the rest. Similarly, the controlling mechanisms employed by the party-state in Mexico are much looser than in most other countries with state-ancillary movements. To be sure, the controls are still there: it is simply that, as Elliot Berg and Jeffrey Butler put it when comparing trade union movements in tropical Africa, 'the levers of control are manipulated more gently and discreetly' in some cases than in others.[52] On the other hand, what this range of control means is that worker protest in countries with state-ancillary movements need not *invariably* involve action outside official union structures.

[51] Schlagheck, *The Political, Economic, and Labor Climate in Mexico*, p. 166.

[52] In Coleman and Rosberg, *Political Parties and National Integration in Tropical Africa*, p. 336; see also Beling, *The Role of Labor in African Nation-building*, pp. 11–13, on degrees of control.

11

PARTY-SURROGATE MOVEMENTS

The party-surrogate type of trade union movement takes one of two forms. On the one hand, major union confederations may *stand in* for a political party in the sense of adopting, in large degree, the functional and structural characteristics of a party without abandoning the concerns and basic structural focus of a trade union body. The other party-surrogate form involves a major confederation, or its constituent unions, acting as the persistently *dominant partner* in a continuing alliance with a party of some political substance (best defined by popular support and/or parliamentary representation).

Trade union movements of this type have been rare. With one exception, they seem to be a feature of the past rather than the present. With the same exception, too, the known cases have all taken the party stand-in form rather than that of dominant partner.

PARTY STAND-INS

Trade union confederations have acted as political parties in circumstances where an appropriate party either did not exist or was unable to operate openly. Colonial regimes in their declining years seem to have provided the most favourable circumstances for the emergence of party-surrogate movements in this form. They have done so by suppressing parties expressive of nationalist movements while permitting (usually subject to close regulation) some form of trade union organization. Trade unionism could thus provide the legal structure within which a national mass movement might be mobilized until open party organization was feasible or political independence was won. Bruce Millen implied that this situation was fairly common in the history of third-world countries,[1] though others have pointed to only two really clear-cut cases—both in Africa, if in the very different cultural settings of Kenya and

[1] See Millen, *The Political Role of Labor in Developing Countries*, p. 9

Tunisia.[2] There would seem to be, however, at least two other possible cases, in Algeria and the Cameroons; and (if without the formal colonial context) versions of the same phenomenon may be detected in South Africa during the 1920s and again in the fifties and sixties.[3] But a more recent and quite unequivocal case occurred on a different continent and in an effectively, though not classically, colonial context: the country is Poland.

Poland

In November 1980 the Polish Supreme Court averted a threatened general strike by overturning the judgment of a lower court. The Supreme Court judges agreed that the rules of an organization formally describing itself as the Independent Self-Governing Trade Union Solidarity (NSZZ *Solidarność*) should be legally recognized —despite the fact that those rules neither disowned the use of strikes nor, apart from a perfunctory reference in an annex, acknowledged the guiding role of the country's ruling party, the Polish United Workers' party (PUWP). Earlier, moreover, a number of other trade unions, with long-standing registration, had also been busy changing their legal names to include the magic words, 'independent self-governing'. These unions were components of the Central Council of Trade Unions (CRZZ) which had, in line with the Soviet model (see Chapter 10), monopolized the Polish trade union scene up to a few weeks before. As it happened, their sudden conversion to verbal independence failed to stem the catastrophic drop in their membership as workers flooded to join the new organization, best known simply as Solidarity. It is thought that, before the year was out, the membership of Solidarity amounted to something between eight and ten million, out of a total population of thirty-six million.

This remarkable combination of events, in a heartland of the Soviet bloc, had its genesis in a strike that broke out in the Lenin shipyard at Gdańsk on 14 August 1980. It was not the first time that Polish workers had struck since the Second World War and the

[2] See Damachi *et al.*, *Industrial Relations in Africa*, p. 203; Davies, *African Trade Unions*, p. 96; Coleman and Rosberg, *Political Parties and National Integration in Tropical Africa*, p. 347.

[3] See Meynaud and Bey, *Trade Unionism in Africa*, pp. 75, 77; Du Toit, *South African Trade Unions*, pp. 35, 41; Feit, *Workers Without Weapons*, pp. 162–3; Luckhardt and Wall, *Organise or Starve*! p. 404.

Soviet occupation. They had done so, most notably and to greatest effect, in 1956 and again, initially in protest against rising food prices, in 1970 and 1976. In 1970, as fourteen years before at Poznań, striking Polish workers were shot down in the Baltic ports of Gdańsk, Gdynia, and Szczecin; in 1976 imprisonment and sackings were the order of the day. Nor was Solidarity the first attempt to organize workers outside the framework of the official unions comprising the CRZZ. During 1978–9 undercover groups claiming to be founding committees of 'Free Trade Unions' were active in the inland cities of Katowice and Radom, as well as in Gdańsk and Szczecin, on the coast.

The strike in the Lenin shipyard triggered further strikes throughout the Gdańsk region and in other industrial centres. Within two days, too, the Gdańsk strikers had set another pattern that others were to follow by forming an Inter-Factory Strike Committee as a regional co-ordinating body. It was this committee, led by Lech Wałęsa, which negotiated with government representatives and on 31 August signed the Gdańsk agreement that signalled the government's formal acceptance of an independent trade union body. A similar but briefer agreement was signed in Szczecin a day earlier.

Independent trade unionism had not been the precipitating issue in the Lenin shipyard strike: that was either, or both (the accounts of participants vary),[4] a wage claim inspired by the recent government-decreed rise in food prices, or the sacking of Anna Walentynowicz, an independent-minded crane-driver. But both these issues were quickly overtaken by others drawn from the reservoir of workers' grievances. In all, twenty-one claims were dealt with in the Gdańsk agreement. The right to form independent trade unions and the right to strike had pride of place. A fortnight after the agreement had been signed, representatives of the Inter-Factory Strike Committees from 33 regions met in Gdańsk to create the national organs of Solidarity; two months later, as we have seen, NSZZ *Solidarność* emerged as the legally registered competitor of the official CRZZ.

Solidarity's legal life was to be a short one. It had time for only one annual congress, which met in September and October 1981. In the meantime it had supported, with strike action, the formation

[4] See esp. MacShane, *Solidarity*, App. 2, pp. 147–50.

and legal recognition of Rural Solidarity, involving the organization of farmers outside the framework of the official rural bodies controlled, like the CRZZ, by the PUWP. Throughout this time, the tension was high, maintained in particular by military manœuvres on the part of Warsaw Pact forces along Poland's borders, hurried consultations between Polish and Soviet officials, and constant reminders about the Soviet invasions of Hungary and Czechoslovakia. The axe fell on 13 December 1981, just sixteen months after the strike started in the Lenin shipyard. Martial law was declared. Solidarity was banned, although not formally dissolved until late the following year, and the arrests began. Solidarity has survived since then only as an underground organization (though it is still acknowledged by the International Confederation of Free Trade Unions as an affiliate; and, beginning in 1987, numerous applications for legal registration have been made by Solidarity local unions, and pursued up the the Supreme Court—so far unsuccessfully). CRZZ, once again, formally has the field to itself.

During its brief legal life, Solidarity displayed characteristics which are highly unusual in the world of trade union confederations. On the structural side, Solidarity's basic unit was the plant or enterprise organization. Dues were collected at this level (the check-off operated for Solidarity, as for the old official unions), and decisions were made about the disbursement of loan and welfare funds. This was nothing out of the ordinary: Solidarity bodies had simply taken over from the parallel organs of the official unions. It was the next level of Solidarity's structure that was different. The intermediate tier between plant and national levels was defined, not in the usual industrial or occupational terms, but geographically. In other words, reflecting the form of organization initiated by the Inter-Factory Strike Committee at Gdańsk, Solidarity structured itself in accordance with the regions into which Poland was divided for the purposes of government and PUWP administration; and, at this level, it purported to represent all employees, irrespective of industry or occupation. This structural oddity, which resembles the 'general', 'omnibus' or 'all-house' unions often figuring in the early history of colonial trade unionism,[5] was the outcome of a deliberate choice. The industrial divisions of the old official trade unions, so

[5] See Millen, *The Political Role of Labor in Developing Countries*, pp. 21–2; Damachi *et al.*, *Industrial Relations in Africa*, p. 202; Martin, 'Tribesmen into Trade Unionists', p. 143.

the argument ran, had only served to divide and weaken them in their dealings with the officials at the regional level of government.

There was, too, another unusual aspect of Solidarity's structure. Like its official counterpart, Solidarity followed the Soviet model in its definition of those eligible for membership, and thus enrolled managerial staff along with all other state employees. In stark contrast to that model, however, its constitution categorically excluded from holding union office all managerial, supervisory, and administrative employees—as well as PUWP officials.

Solidarity's policy aims have been depicted as evolving through three stages, labelled Solidarity I, II, and III.[6] The twenty-one points covered by the Gdańsk agreement of August 1980 largely involved wage and welfare matters, but there were also broader claims to free speech, the lifting of censorship, the freeing of political prisoners and, as already mentioned, the right to strike and the right to form independent trade unions. Given the Leninist conception of the trade unions in a socialist state as indispensable 'transmission belts' between the Communist party and the working masses, the demand for independent trade union organization was a supremely political claim. That was Solidarity I.

Solidarity II centred on the broader claim for worker self-management in industry. The government conceded this claim in legislation (but in practice not at all) with the passage of the Self-Management Law of September 1981. Solidarity III emerged as an outcome of the one and only annual congress, held in Gdańsk where it all began. The earlier claims, in a Communist state, had been highly political in their implications. The programme which the 1981 congress adopted was political in a much more expansive sense. The formal demands now included, among other things, free nomination (that is, not conditional on PUWP approval) in forthcoming local elections; the formation of a new house of parliament to comprise representatives of 'social and economic groups'; a 'social council for the national economy', staffed by experts and equipped with substantial administrative powers; and, 'ultimately, free elections to Parliament'.[7]

Thus by late 1981 Solidarity, in both its structure and its policy programme, had more and more come to look like a political party. From the start, indeed, there had been an ambivalence about

[6] Touraine *et al.*, *Solidarity*, pp. 97–8. [7] Ibid. 98.

Solidarity's credentials as a trade union body pure and simple. It had emerged on a swelling tide of nationalist and anti-Soviet sentiment, which had been profoundly influenced by the triumphal tour of the 'Polish Pope', John Paul II, in June 1979. All the symbols were there against a wall of the room used by the Gdańsk Inter-Factory Strike Committee in August 1980: a life-size plaster statue of Lenin, a crucifix, and a Polish flag. The specifics were there in demands made at the start, but later dropped, by the strikers at Gdańsk ('free elections') and at Szczecin ('independent political parties').[8] The recognition of the ambivalence was there in the undertaking which government negotiators secured in the Gdańsk agreement: 'The new trade unions . . . do not intend to play the role of a political party.'[9]

Solidarity's leaders, for the most part, continued to insist that 'Solidarity is a trade union, not a political movement'.[10] But both within the organization and outside it, the question was constantly pressed, in one form or another—and sometimes as bluntly as at a mass meeting in Szczecin: 'When would the union become a political party?'[11] For few believed, in the circumstances of the time, that a free trade union could confine itself to industrial matters. The leadership recognized this, but was anxious to deny the party label. Thus Lech Wałęsa, asked to explain 'what Solidarity was', is reported to have replied that 'it was more than a trade union, and not the same thing as a political party, but rather a social movement'.[12]

Solidarity's expressed concerns made it easy for opponents to charge it with exceeding a trade union brief. As early as February 1981, Stanisław Kania, first secretary of the PUWP, publicly accused it of 'behaving like a political party'.[13] The congress of September–October gave credence to this view, not merely by the policy programme it adopted, but also by an open letter which it addressed to Poles outside the country, and in which it asserted that Solidarity was 'not only a trade union, but also a social movement of thinking citizens wishing to work for Poland's independence'.[14]

[8] Touraine *et al.*, *Solidarity*, p. 37; Ascherson, *The Polish August*, p. 169.
[9] Quoted in Ascherson, *The Polish August*, p. 284.
[10] MacShane, *Solidarity*, p. 126.
[11] Ascherson, *The Polish August*, p. 205.
[12] Touraine *et al.*, *Solidarity*, pp. 55–6.
[13] Ascherson, *The Polish August*, p. 261.
[14] Quoted in Touraine *et al.*, *Solidarity*, p. 140.

It might well be true that when Solidarity called for free elections, 'its aim was not to become a political party and win those elections, but to abolish itself in the multiplicity of political parties or to return to the more limited function of a trade union'.[15] Nevertheless, whatever the intentions of its leaders (and they were plainly mixed), the reality was that in the meantime Solidarity was sucked inexorably into a deepening political role by a groundswell of nationalist feeling which sought expression in circumstances where alternative channels of popular representation were unavailable.

DOMINANT PARTNERS

There is a common belief, in many liberal democracies, that a political party associated with the trade unions is persistently dominated by them. It is a belief that acquires added colour when the association involves formal affiliation. Closer examination, however, reveals that trade union domination of the party is usually either transitory and irregular or is non-existent—at least in terms of the conventional picture depicting a trade union monolith lording it over the party's politicians and non-union membership (see Chapter 13). Nevertheless, there is one country, and two parties, in relation to which the proposition about continuing trade union domination appears to have some substance. This case of a party-surrogate trade union movement comes from Asia: the country is Japan.

Japan

Japanese trade unionism has caught the attention of foreign observers primarily because of its unusual structural characteristics. In the first place, trade union organization there is almost wholly confined to public employment and to large-scale concerns in the private sector. Secondly, within these recruitment areas 'temporary' employees, as in China (see Chapter 10), are for the most part excluded from union membership, which tends to be reserved for employees who have an assurance of lifetime tenure after joining a firm or agency on completing their formal education. Finally, although there is a smattering of craft, industrial, and general unions, something like 90 per cent of trade union members belong

[15] Ibid. 100.

to the thousands of 'enterprise unions' which operate in both the public and the private sector. The prevalence of this structural form, while not as singular as often assumed (it is standard in Pakistan, Bangladesh, Thailand, South Korea, and a number of Latin American countries;[16] and is common in Malaysia), is unique in the case of countries with more advanced market economies.

Enterprise unions may be unions limited to a firm with one plant or they may be unions with branches in the various plants of a larger enterprise. In either case, they are officially known as 'unit unions' (and total well over 30,000); but they become 'basic unit unions' (and more than double in number) when the branches of the second type are included in the count on the ground of their capacity for independent activity. The Japanese enterprise unions are unusual in that, as well as manual workers, they typically recruit white-collar employees up to the lower levels of management, including professionals. Their full-time officials are drawn from their own membership (usually the white-collar segment) and retain their career entitlements in the enterprise, to which they usually return. It is the enterprise union which plays the dominant role on the trade union side in the negotiation of pay and working conditions, and in the handling of industrial and welfare matters in general.

Direct concern with specifically industrial matters tends to decline sharply as the level of trade union organization rises. Above the enterprise union there are industry federations, linking enterprise unions within the one industry, and national confederations. Single-plant enterprise unions affiliate directly or indirectly at both of these levels, but do so usually for political rather than industrial reasons. Neither the industry federations (for the greater part) nor the national confederations have much to do with the detail of collective bargaining negotiations, though some of the latter have had a hand in setting parameters through annual campaigns ('the Spring Offensive' or Shuntō) on the wage issue. But the principal concerns of both are usually political in character. This is especially so in the case of the confederations.

At the confederal level, the Japanese trade union movement looks like a party-ancillary movement, with two confederations linked to different political parties, and others defining themselves as

[16] See esp. Cordova, *Industrial Relations in Latin America*, pp. 30–1.

politically independent. The competitive element is much more evident at the upper levels than the lower, where it seems that, as a rule, an enterprise union has the field to itself. At the same time, breakaways are not uncommon, and the existence within the one enterprise of two, and sometimes three, unions associated with different confederations is not unheard of. But what takes the Japanese trade union movement out of the party-ancillary category is the peculiar relationship that exists between each of two major confederations and the party with which it is associated.

A Japanese Federation of Labour (Sōdōmei), founded in 1919, led a precarious existence during the inter-war years until it was suppressed in 1940. When it was re-formed early in 1946, there were two major left-wing parties in existence: the new Japan Socialist party and the old Japan Communist party. The leadership of Sōdōmei was associated principally with the controlling right-wing faction of the Socialist party. Later in 1946, the Communists and left-wing elements of the Socialist party formed a rival confederation, the National Congress of Industrial Unions (Sambetsu). Following a swing to the left in both Sōdōmei and the Socialist party, a new and purportedly anti-Communist confederation emerged in 1950, with the backing of the Occupation authorities. This, the General Council of Trade Unions of Japan (Sōhyō), attracted most of the older affiliates of Sōdōmei, together with some affiliates of the recently disbanded Sambetsu.

The Socialist party split in 1951 over the Japanese Peace Treaty, its right wing breaking away to form the Democratic Socialist party—a split that was temporarily healed four years later, but reaffirmed in 1960. Sōhyō followed suit in 1954, when right-wing unions broke away to join those still linked to the old Sōdōmei to form a new confederation (initially known as Zenrō) which by 1964 had settled down as the Japan Trade Union Congress (Dōmei). Two other confederations emerged during the 1950s: the Federation of Independent Unions (Chūritsu Rōren), and the National Federation of Industrial Organizations (Shinsanbetsu). Both claim to be independent of political parties.

These four confederations together cover something under two-thirds of all trade union members: Sōhyō accounting for about 34 per cent (or 4.5 million) in 1984, Dōmei for some 17 per cent, Chūritsu Rōren for 11 per cent, and Shinsanbetsu for less than one per cent.[17] Sōhyō's membership is dominated by unions with members employed either in public enterprises or directly by local and central government. Unlike private sector employees, both types of public employees are legally denied the right to strike; and direct government employees are denied formal collective bargaining rights as well. Their unions lean strongly toward political forms of action. The other three confederations are concerned primarily with private industry. This has not, however, obviated overlapping organization and other symptoms of rival unionism between them and Sōhyō, including official encouragement of breakaways. Sōhyō's relations with Dōmei have always been especially competitive.

Despite their ideological differences, the four confederations displayed a remarkable readiness to co-ordinate their annual, shuntō, wage claims from 1976. Since then, moreover, there have been moves among their constituents towards greater organizational unity, culminating in 1982 in the formation of the Japanese Private Sector Trade Union Council (Zenmin Rōkyō)—the 'Council' being altered to 'Confederation', and the shortened name to Rengō, in late 1987. In both its forms, Rengō has professed the ambition of bringing the whole Japanese trade union movement within the embrace of one organization; but, as its name indicates, has so far confined its affiliations to private sector unions. It aims to avoid special links with particular political parties, without restricting individual affiliates in this respect. Its strongest opposition comes from the extreme left, and especially from elements associated with Sōhyō. It appears to have attracted affiliates (without requiring them to sever the earlier connection) from all the older confederations, as well as from among the large group of non-affiliated unions. Its coverage of trade union members, at some 37 per cent of the total, is greater than Sōhyō's. On the other hand, at least up to 1987, its staffing and financial resources were much slighter, with the result that it operated less as a full-blooded confederation than as 'a loose coordinating council or forum for joint discussions and planning'.[18] It remains to be seen whether the new name reflects a change in this respect.

In the meantime, Sōhyō and Dōmei, and most of their affiliated

[17] Bamber and Lansbury, *International and Comparative Industrial Relations*, p. 218.

[18] Juris *et al.*, *Industrial Relations in a Decade of Economic Change*, p. 291.

unions, have each maintained—as for many years past—close ties with a political party: the Socialist party and the Democratic Socialist party, respectively. And, in each case, the crucial point is that the confederation and its associated unions have consistently played the role of dominant partner in the relationship with the party.

There are two factors of particular significance in this highly unusual situation. One is that the two parties have come to form what amounts to a permanent opposition in the Japanese legislature, the Diet. Neither has had the taste of government office on its tongue since 1947–8, when the Socialist party formed part of a coalition; and since 1955 the Socialist party has been the major opposition party in the Diet, confronting an uninterrupted sequence of governments controlled by the Liberal Democratic party.

The second factor is that, like the ruling party, neither the Socialist party nor the Democratic Socialist party is a mass party. In the case of the Socialist party, this is reflected in the fact that despite its very much larger Diet representation in 1980 (107 seats to 29), its membership was less than 40,000 as compared with the 300,000 of the Communist party. At the local level, indeed, 'organizers' seem to comprise most of the party's active membership. Otherwise, local party organization depends largely upon the Sōhyō unions, 'whose wide distribution gives the Socialists alone among the opposition parties a national base'.[19] The trade unions thus provide 'a kind of substitute organization'; and their concerns, as a result, 'tend to dominate party policy-making'.[20] Sōhyō and its affiliated unions are also the main source of Socialist party finance, for both campaign and general expenses, and provide election campaign workers and other facilities. Above all, a great many of the party's candidates in Diet elections are customarily nominated directly by friendly union bodies in the party's selection process. It is this 'continued dependence upon trade unions both for its supply of electoral candidates and for much of the logistics of local organization' which Stockwin has described as the Socialist party's 'Achilles heel'.[21] Another observer judged Sōhyō's influence within it to be such that 'the party might [almost] be considered merely as a pressure group in disguise'.[22] And yet another has described Sōhyō

[19] Macdougall, *Political Leadership in Contemporary Japan*, p. 57.
[20] Stockwin, *Japan*, p. 180. [21] Ibid. 176.
[22] Willey, 'Pressure Group Politics', p. 707.

as 'a powerful force in holding the party together', citing an incident in which Sōhyō prevented a major party split by threatening to 'withhold financial and organizational support from . . . discontented Socialist Diet members'.[23] Sōhyō's influence is reflected, in a different way, in the persistently doctrinaire character of the Socialist party's policies. In this, the party is a 'ready instrument for the unions to use in their political programmes'.[24] The Sōhyō unions (given the relatively narrow ideological focus and identification of Sōhyō itself) have ensured that 'ideological extremism' has remained the hallmark of the Socialist party, despite attempts by minority factions to expand its policy lines in the hope of broadening its electoral appeal.[25]

As for the corresponding relationship between Dōmei and the Democratic Socialist party, an early study of Dōmei's antecedent, Zenrō, concluded that its connection with that party was 'closer than Sōhyō's with the JSP'.[26] Alice Cook later underlined the 'extreme dependence' of the Democratic Socialist party on the Dōmei unions for financial and organizational support, a dependence on which its 'very existence' turned, so that 'the party without the unions is unthinkable'.[27] It also follows the Socialist party's practice of allowing sympathetic unions to make direct nominations in ballots for Diet candidates. On the other hand, Dōmei's influence on party policy has been far less restricting than Sōhyō's, as might be expected in the case of union leaders with reformist rather than revolutionary leanings. Their moderation was reflected in the Democratic Socialist party's readiness, in the late 1970s, to contemplate the possibility (which did not, in the end, eventuate) of a junior role in a coalition government with the Liberal Democratic party.

SUMMATION: DEGREES OF SUBSTITUTION

The range of the party-surrogate category is encapsulated in the two quite distinct forms which this type of trade union movement may take. On the one hand, there is the stand-in form. In this case, given the legal or actual absence of an organization that would

[23] Macdougall, *Political Leadership in Contemporary Japan*, p. 70.
[24] Cook, *An Introduction to Japanese Trade Unionism*, p. 111.
[25] Stockwin, *Japan*, p. 166.
[26] Quoted in Cook, *An Introduction to Japanese Trade Unionism*, p. 150.
[27] Ibid. 149, 151, 171.

normally be described (and would describe itself) as a political party, the trade union confederation *is* the party. On the other hand, there is the dominant-partner form. In this case the trade union confederation and its unions are associated with an organizationally distinct political party which they largely and persistently control in terms of organization, policy, and parliamentary representation.

12

THE STATE-SURROGATE CATEGORY

Only one trade union movement has come at all close to the role of state-surrogate, in the sense of either continuously dominating or standing in for a government. It is the trade union movement of Israel.

Israel

The single Israeli trade union confederation, the General Federation of Labour in Israel, is familiarly known as the Histadrut, a Hebrew word meaning simply 'organization'. The Histadrut functions as a trade union confederation in that it incorporates a number of trade unions and is concerned with the negotiation of pay and working conditions; it also, on occasion, authorizes strike action. But the Histadrut is more than just a trade union confederation, a fact that is demonstrated by the nature of its membership. It is technically possible to become a member of one of the Histadrut's constituent trade unions without becoming a member of the Histadrut itself. Usually, however, employees first join the Histadrut directly, and are then allocated to the appropriate union; their subscriptions, accordingly, are paid to the Histadrut which, in turn, distributes a share to the union. But eligibility to join the Histadrut itself is not confined to employees, or to members of Histadrut unions. Full membership is also open to housewives, members of co-operatives, and self-employed tradesmen and professionals. It claims to cover 90 per cent of the country's 'working population' (including Arabs, who have been eligible for full membership since 1959).[1]

The great bulk of those Histadrut members who are also unionists are distributed among some forty national trade unions integrally linked to it; a small minority belong to independent unions, mainly of white-collar and professional employees. Most of

[1] *Histadrut*, p. 5. This excludes Arabs who both live and work in the occupied territories of the West Bank and the Gaza Strip: they are organized in their own unions and, on the West Bank, their own confederation.

the Histadrut unions were created by the Histadrut itself, but some were not. They are organized variously on industrial, occupational, or craft lines. Below the national level there is a pattern of local trade union organizers, regional labour councils, and works committees. This structure, from national unions to plant organization (but excluding the labour councils which come under another department), is officially administered by the Trade Union Department of the Histadrut. The department itself is formally subject to the Histadrut's central organs.

The principal legislative body of the Histadrut, its convention, is regularly elected by way of procedures which parallel the electoral process relating to Israel's parliament, the Knesset. All the main political parties are officially involved in Histadrut general elections. As for the Knesset, they each submit a list of candidates in nation-wide Histadrut elections based on single-vote proportional representation, seats in the convention being allocated to each party in accordance with the proportion of the total national vote favouring its list. The same proportions are also used to distribute seats on other bodies of decreasing size that meet more frequently than the convention: the general council, the executive committee, and the central executive committee. Administrative positions are allocated in the same terms. Separate elections but the same process, with party participation, applies as well as in the case of each of the regional labour councils and the national trade unions.

The unusual character of the Histadrut is further reflected in the diversity of its concerns. Acting either directly or through agencies with which it is formally associated, it is involved in a range of activities that is unmatched in the case of any other national confederation. In the first place, it is concerned with the negotiation and settlement of wages and working conditions. Secondly, it is heavily involved in the provision of social welfare facilities, including nation-wide medical and hospital services; insurance funds which substantially supplement state benefits relating to old age, sickness and accident; loan schemes, old-age homes, and orphanages. Thirdly, its cultural, educational, and recreational concerns are reflected in its responsibility for adult and vocational education; and in its ownership of two daily newspapers and a number of periodicals, a publishing house and bookstores, a chain of sports clubs, and a theatre. Finally, and most striking, there is the Histadrut's involvement with economic enterprises—agriculture and agricultural marketing; irrigation and water supply; banking and general insurance; airlines, shipping, and land transport; building and construction; general manufacturing and processing; and retail department stores.

The significance of the Histadrut's wide-ranging concerns may be indicated in a number of ways. Those it employs, both directly and indirectly, amount to about a quarter of all Israeli employees; and it is the country's single largest industrial employer. Its health and medical services, Kupat Holim, cater for more than 75 per cent of the total population. Its construction contracting company, Solel Boneh, is the largest in the Middle East. Moreover, as is obvious, many of its activities are in other societies either wholly or largely the province of the state.

The overlap with state-like activities was once even more pronounced. Before the creation of the state of Israel, the Histadrut was the sole source of insurance against the hazards of old age, sickness, and accidents for the Jewish community in Palestine. It also developed and administered the main non-religious school system and a network of labour exchanges, both of which have since been taken over by the state. For a time, too, the Histadrut even discharged the most characteristically state-like function of all: it ran an army.

The explanation of this extraordinary range of concerns is to be found in the Histadrut's origins and the circumstances of its early existence. It was founded in 1920. The drive to establish a Jewish homeland in Palestine had acquired new weight as a result of the Balfour Declaration and the decision, formally effected in 1922, to place Palestine under a British Mandate. The Histadrut was formed by the political parties of the Labour–Zionist movement. Its antecedents were some embryonic trade union organizations, mainly sponsored by parties, a Jewish health service and insurance fund dating from 1911, and a co-operative enterprise formed during the First World War to handle the purchase and distribution of supplies to Jewish settlements. The Histadrut assumed the functions of these bodies and quickly acquired others. Its entry into the field of industrial enterprise, its organization of labour exchanges, its creation of a school network, and its promotion of cultural activities all date from this time. And among its earliest tasks was that of creating a militia force, the Haganah ('defence' in Hebrew), which was to prove 'the major forerunner of the Israeli

Army'.[2] This, it should be said, is not quite unique. Trade unions in other countries have founded and funded militia forces on a strictly temporary basis, usually in revolutionary situations (Lenin, for one, regarded this as a central trade union function). But there is at least one closer, because more lasting, parallel. In Bolivia, between 1952 and 1958, trade union militias replaced the standing army after that had been disbanded by the government; the trade union confederation, in addition, was given a formal right of veto in relation to all important government legislation.[3]

Almost from the start, the Histadrut had a special place in the affairs of the Jewish community in Palestine. There were some other notable inter-party bodies, including the Jewish Agency (established in 1922 to advise the mandatory government), the National Council of Jews in Palestine, and the Municipal Council of Tel Aviv, an all-Jewish city. Of these, the Jewish Agency was particularly significant. But it fell to the Histadrut to provide the principal means by which the expanding Jewish community, during the time of the British mandate, established the modern social and economic infrastructure inherited by the state of Israel. To its founders, the Histadrut was 'the socialist state on the way'.[4] By the end of the 1930s, as Howard Sachar put it, 'it had become much more than a powerful institution in Jewish Palestine'—to the greater part of the Jewish community, 'the Histadrut was all but synonymous with Jewish Palestine itself'.[5] Even after the creation of the state of Israel, there was for some time still, in many quarters, 'a tendency to regard the Histadrut as a kind of workers' state which almost autonomously supplies all the needs of the working community'.[6] Ferdynand Zweig was to make the point in a different way when he wrote of the Histadrut that 'it was actually the forerunner of the Jewish State, covering all aspects of Jewish social, economic and cultural life'.[7] And, according to Zweig, it has not altogether lost the characteristics which prompted that remark: 'The Kingdom of the Histadrut, as it is often called, is still a sort of state within the state'.[8]

[2] Rothenberg, *The Anatomy of the Israeli Army*, p. 23; see also Luttwak and Horowitz, *The Israeli Army*, pp. 8–9.

[3] See Magill, *Labor Unions and Political Socialization: A Case Study of Bolivian Workers*, pp. 26–31.

[4] Shaari, 'Relationship of Histadrut and State', p. 20.

[5] Sachar, *A History of Israel*, p. 159.

[6] Shaari, 'Relationship of Histadrut and State', p. 21.

[7] Zweig, 'The Jewish Trade Union Movement in Israel', p. 166.

[8] Ibid. 162.

SUMMATION: THE HISTADRUT'S SINGULARITY

The Israeli trade union movement, arguably, has affinities of a sort with three other types of movement: the party-ancillary, the state-ancillary, and the autonomous. The party-ancillary affinity is implied in Peter Medding's assertion that 'the political parties control the Histadrut and, to be more exact, Mapai . . . controlled the Histadrut'[9]—Mapai being the party which, either in its own right or in coalition with other socialist parties, has formed the majority grouping within the Histadrut since its foundation. Medding's own elaboration on this theme suggests that his notion of Mapai 'control' needs to be modified in the case of a Histadrut leadership which has both wrenched an explicit acknowledgement of its independence from the Mapai and, more openly, asserted its independence of Mapai governments.[10] On the other hand, the formal electoral competition of opposition parties within the Histadrut, while not at all typical of party-ancillary movements, does find a parallel in the case of Venezuela (see Chapter 9). But, crucially, the Histadrut lacks the party-aligned confederal competition of the Venezuelan case.

As for the state-ancillary category, there is an affinity reflected in the Histadrut's quite strongly centralized structure and in the extent to which its concerns stretch to areas conventionally regarded as governmental. In addition, its leaders have sometimes *sounded* like those of state-ancillary trade union movements: 'Histadrut policy has . . . consistently aimed at increasing output and productivity', and the 'interests of the whole nation strongly influence the formulation of trade union policy'.[11] But despite the rhetoric, the Histadrut has on many occasions demonstrated its independence of government—not least by its preparedness to endorse strike action, including (as in 1985) a general strike against the economic policies of a government with a Mapai prime minister. Further, its functions extend far beyond those permitted to state-ancillary movements, especially in their commercial and industrial ramifications. The state-ancillary model is also belied by the Histadrut's internal political arrangements, involving as it does multi-party competition and representation.

[9] Medding, *Mapai in Israel*, p. 184.

[10] See ibid. 186, 194–5.

[11] Becker (former Histadrut general secretary), 'The Work of the General Federation of Labour in Israel, pp. 442, 449.

In the case of the autonomous category (considered below in Chapter 13), there is a strong structural parallel between the Histadrut and the Austrian national confederation (OGB): they have in common a high degree of organizational centralization and an official multi-party electoral process. Functionally, there is something of a parallel with the main West German confederation (DGB) in relation to social security administration and to commercial and industrial enterprises. But the DGB's concerns are more limited than the Histadrut's in these areas; and, unlike the Histadrut, it has little or nothing to do with collective bargaining and the settlement of pay and working conditions. Nevertheless, of all categories other than state-surrogate, autonomous is the one with which the Histadrut most nearly complies on the basis of its relations with state and party. It has been pushed towards that category since the creation of the state of Israel in 1948, because of the reduction in its functions through transfer to the state. This, of course, implies that the Histadrut's claim to singularity, and accordingly the case for identifying Israeli trade unionism as a state-surrogate movement, is most compelling when couched in historical terms. And that is true. It is also true, however, that the trade union movement of contemporary Israel retains sufficient characteristics, inherited from the period of the British Mandate in Palestine, to mark it out as unequivocally one of a kind.

13

AUTONOMOUS MOVEMENTS

Both the structure and the functions of autonomous trade union movements vary considerably. So, too, do their precise relationships with political parties and with the state. On the other hand, all those relationships share the same negative characteristic: they do not involve (except on an occasional basis) the *domination* of one side by the other.

A round dozen of the countries with autonomous trade union movements—amounting to nearly all such movements—fall into the select grouping of eighteen nations which the World Bank designates 'industrial market economies'.[1] Six of them are English-speaking countries: the United Kingdom, Ireland, the United States, Canada, Australia, and New Zealand. The other six are in Continental Europe: the four Nordic nations (Sweden, Norway, Denmark, Finland), and two of the German-speaking (West Germany and Austria). The trade union movements in these twelve countries are the focus of this chapter.

There seems to be no more than three other substantial trade union movements with a serious claim to the autonomous label. Only one of them, that of Iceland, falls without qualification into the category. Fiji, it is true, had a trade union movement that was clearly autonomous up to 1987; but its position has become uncertain, following two military coups. The likelihood is that it has started to slide into the state-ancillary category. Then there is the curious case of Argentina. From 1955 the Argentine trade union movement—a state-ancillary movement during Juan Perón's dictatorship (1946–55) and closely associated with the Peronist party since then—was the subject of a number of attempts, under both military and elected regimes, to force it back into the state-ancillary mould. Each time, however, it either retained under duress or, eventually, re-emerged with the essential qualities of an autonomous movement. Fairly described as 'the most durable and

[1] The others are Italy, France, Japan, the Netherlands, Belgium, and Switzerland: see the World Bank's annual *World Development Report*, Annex.

independent' of all trade union movements in Latin America,[2] the Argentine movement is also unique among autonomous movements in the extreme volatility of its fortunes. It is perhaps best characterized as a persisting and resilient, but distinctively unstable member of the autonomous category.

STRUCTURE AND FUNCTIONS

In the case of the dozen main autonomous movements, union density rates (unionists as a proportion of all employees) range from above 80 per cent in Sweden to less than 20 per cent in the United States. A majority boast density rates above 50 per cent: in *rough* rank order (precise comparisons in this respect being extremely hazardous)[3] they are Sweden, Austria, Finland, Norway, New Zealand, Denmark, Australia, and Ireland. Two (United Kingdom and West Germany) have rates in the forties; while Canada, at about 30 per cent, comes closest to the low American figure.

The trade unions containing these memberships tend to a greater variety, in the basis on which they recruit members, than their counterparts in party-and state-ancillary movements. The diversity is particularly marked in the English-speaking cases, which each involve a mix that includes unions organized in terms of specific skills or occupational categories, in terms of a specific industry, or along lines (the 'general' or 'conglomerate' union) with no discernible rationale other than organizational opportunism. In the Continental cases, the principle of industry-based organization predominates among manual workers, except in Denmark where craft unions are a major factor; the occupational principle otherwise finds expression mainly in relation to the organization of white-collar and professional employees. The structural picture is complicated in Canada and in Ireland by the presence of 'international' unions with headquarters in another country, the United States and Britain, respectively. The branches of such unions now account for something over 40 per cent (the corresponding figure was more than 70 per cent in 1966) of all trade union members in

[2] Wynia, *Argentina*, p. 163.

[3] See, esp. ILO, *Collective Bargaining in Industrialised Market Economies*, p. 31; Bamber and Lansbury, *International and Comparative Industrial Relations*, pp. 256–7; von Beyme, *Challenge to Power*, pp. 75–6.

Canada, and for 14 per cent in Ireland. In Ireland, there is the further complication that a number of Irish trade unions themselves operate internationally by way of either a head office or branches in Northern Ireland.

There is a similar variety in the raw number of trade unions (including national or federal unions and independent state or local unions), which tends to be much greater in the English-speaking countries. In those cases, with the exception of Ireland, the range is from something over 200 separate unions to a little under 500—with Canada at the lower end of the scale, the United States and the United Kingdom at the higher end, and Australia and New Zealand in the middle. In Ireland and in each of the Continental countries the number of separate unions is strikingly lower—and, indeed, appears to exceed 100 only in the case of Finland. By far the lowest is the 17 unions in which the trade unionists of Austria are grouped. West Germany would come close to this figure, with Sweden not far behind, if only their industrially-based manual workers' organizations were counted; but white-collar and professional unions, as in the other cases, greatly swell the numbers.

In every autonomous trade union movement there is one national confederation which either stands alone or, if it does not, at least overshadows the field. Each of them, through its associated unions, covers a decisive majority of the unionized work force. In five cases (the rank order is Austria, United Kingdom, Ireland, Australia, and West Germany), the proportion covered is well over 80 per cent; in others it drops into the seventies (United States, New Zealand, Denmark, Norway, and Canada) and through to the sixties (Finland and Sweden).

Four confederations stand alone in their national context. They include the oldest of all the world's national confederations, the British Trades Union Congress (TUC), together with the American Federation of Labor and Congress of Industrial Organizations (AFL–CIO), the Austrian Trade Union Federation (OGB) and the New Zealand Council of Trade Unions (NZCTU). But while these four have no confederal competitors, like their state-ancillary counterparts, they differ from the latter in that, among other things, none of them encloses every trade union in the country. The OGB in Austria comes closest to doing so; only one union, covering rural workers, escapes its net. In the case of the other three solo confederations, the number and the strength of independent unions are greatest in the case of the AFL–CIO.

The principal confederation in every other autonomous trade union movement faces one or more confederal competitors. With minor exceptions affecting the German, Swedish, and Canadian movements, the lines of division between confederations are drawn in *occupational* terms rather than in the political terms characteristic of party-ancillary movements (though it is true that the occupational distinction often has political overtones).[4] Thus the single confederation which stands alongside the Irish Congress of Trades Unions (ICTU) and the Australian Council of Trade Unions (ACTU), respectively, expressly concerns itself with professional employees. Similarly, the three confederal bodies confronting the Norwegian Trade Union Confederation (LO) and the Central Organization of Finnish Trade Unions (SAK) differentiate themselves in terms of white-collar, technical, and professional employees. The same three categories, plus government employees, distinguish the four confederal rivals of the Danish Trade Union Confederation (LO).

Like their Norwegian and Finnish counterparts, both the West German Trade Union Federation (DGB) and the Swedish Trade Union Confederation (LO) each face three rivals—but only two of them are defined in occupational terms (white-collar and government service in West Germany, white-collar and professional in Sweden). The third defines itself politically. In the West German case, the Christian Trade Union Federation, professing to be anti-socialist and incorporating specifically manual, white-collar and government service unions, accounts for something under three per cent of all union members. For its part, the Central Organization of Swedish Workers, professing a syndicalist position and confined to manual unions, covers no more than half of one per cent of all

[4] The history of the American trade union movement provides a unique variant on this theme. The twenty-year division between the old American Federation of Labor and the breakaway Congress of Industrial Organizations, before they formed the AFL–CIO in 1955, stemmed from the conflict between the craft and the industrial principles of organization: see Galenson, *The CIO Challenge to the AFL*, ch. 1; and Tawney, *The American Labour Movement*, pp. 23–30. The occupational nature of this division differed from the cases mentioned below: it involved, in effect a distinction between one confederation of manual unions dominated by skilled workers (AFL) and another of manual unions dominated by unskilled workers (CIO).

trade unionists in Sweden. In Canada, too, the dominant Canadian Labour Congress (CLC) faces one minority confederation which defines itself politically, together with another which does not (the latter consists of some international union branches which recently broke away from the CLC). The former, the Confederation of National Trade Unions, is in fact confined wholly to the French-speaking province of Quebec, and is a reflex of French-Canadian separatism: although it covers almost 50 per cent of all trade union members in Quebec, it accounts for less than six per cent nationally. Thus in the West German, Swedish, and Canadian cases, there is a minor party-ancillary element in an otherwise unambiguously autonomous type of trade union movement.

There are enormous variations in the authority which the solo or dominant confederations wield over their constitutent trade unions. By far the most highly centralized is the Austrian OGB, which is unique among autonomous trade union movements in the degree to which it resembles the centralized confederal structure common among party and, especially, state-ancillary movements. Like the German DGB, it is a post-war blueprint creation. Its sixteen trade unions, all industry-based apart from one covering white-collar employees in private industry, are not affiliated like the unions of the other solo or dominant confederations. Instead, like the unions associated with the Israeli Histadrut (see Chapter 12), they are branches or departments: they have no legal status apart from the OGB. Their constitutions and finances are formally, and effectively, controlled by the OGB's central organs, which also appoint their secretaries and other staff, decide their policies, and approve any agreements they make with employers. The American AFL–CIO stands at the opposite extreme. It is an untidy collection of more than a hundred craft and industrial unions which choose to remain affiliated to it. Their constitutions, finances, staffing, policies, and industrial agreements are (with limited exceptions relating to Communism, corruption, and discrimination) entirely beyond its formal reach, and almost as completely beyond its effective influence.

Perhaps the litmus test of confederal authority between these extremes is a confederation's involvement in the settlement of pay and working conditions, because these are the issues on which the concerns of trade unions in autonomous movements are perennially focused. In the nature of the case, no great precision is possible.

Nevertheless, it can be said, for example, that the Norwegian LO seems to stand nearest to the Austrian OGB in terms of authoritative involvement in its affiliates' collective bargaining concerns. The Swedish LO, once close on its heels, has since 1984 fallen further down the scale; but together with the Finnish SAK, the Danish LO, the Australian ACTU and, probably, the New Zealand NZCTU, still has more affinity to the OGB than to the AFL–CIO in this respect. Decisively in the lower levels of the scale are the British TUC, the Irish ICTU and, closest of all to the AFL–CIO's position, the Canadian CLC and the German DGB.

But if the DGB is low on the authority scale, as measured by its lack of a collective bargaining function, it far outstrips the confederations of other autonomous movements in another functional aspect—its involvement in business enterprises. Up to 1986, when it reportedly began to dispose of some of these assets, the DGB owned West Germany's largest life insurance company, one of the three biggest travel companies, the fourth largest commercial bank, a major building society, a housing construction company, and the largest property-development company in Western Europe. The range and scale of its business interests are surpassed only by those of the Histadrut in Israel (see Chapter 12). Within the autonomous category proper, it appears that no more than two other confederations even remotely approach the DGB's business involvement. The Australian ACTU, in combination with private enterprise, runs a petrol retailing company and a leading travel agency, and has begun to move into the field of superannuation and financial services. The Swedish LO has a range of business interests which is wider (they include newspapers, property development, printing and data-processing firms, and holiday homes in Italy), but still well short of the DGB's. The AFL–CIO, although otherwise much less involved in the business area, has recently entered it in an unusual—but perhaps distinctly American—way by establishing its own credit-card and product-endorsement enterprise.

RELATIONS WITH POLITICAL PARTIES

An odd feature of the autonomous category is that *formal* linkages between political parties and trade unions are common. Typically, they do not involve a major confederation, but take the form of other union bodies affiliating officially to a party, and thus being entitled to direct representation in certain party organs. Altogether seven of the main autonomous trade union movements are associated with parties to which trade unions, but not all trade unions, are connected by way of formal affiliation. The singularity of this kind of party–union relationship is demonstrated by the fact that, in the case of all other types of trade union movement, there are probably no more than five other contemporary instances: Mexico (see Chapter 10) and Peru, in Latin America; and, in Africa, Tanzania (see Chapter 10), Senegal, and Malawi. All except the Peruvian case (party-ancillary) fall into the state-ancillary category. There are, however, other historical cases, as in the party-ancillary movements of Germany (before 1933), of Austria (before 1934) and, before the mid-1940s, of Belgium, the Netherlands, and Switzerland. In Africa also, there was a parallel in Nigeria, before independence. Elsewhere, it is uncertain that a union–party connection of this kind, created in 1985, will long survive the 1987 military coups in Fiji.

As to the twelve main autonomous movements, the party-affiliation linkage occurs in five of the six English-speaking countries (the exception being the United States), but in only two of the Continental cases, Sweden and Norway. The parties involved formally describe themselves as 'Labour' parties except in Sweden (Social Democratic party) and Canada (New Democratic party). They all have in common, as well as the principle of trade union affiliation, the nature of their origins. For they were all founded by trade unions. More than this, the two Continental parties originally filled a double role as both party and trade union confederation (the Swedish Social Democratic party from 1889 to 1898, and the Norwegian Labour party from 1887 to 1899). In contrast, the Irish Labour party functioned as a component, first, of the 'Irish Trades Union Congress and Labour Party' from 1912 to 1918, and then of the 'Irish Labour Party and Trades Union Congress' until 1930.

Most of these seven 'Labour parties' are unquestionably major parties. The exceptions are the Irish Labour party and the New Democratic party of Canada. Unlike their British, Australian, New Zealand, Swedish, and Norwegian counterparts, they have never had control of national government in their own right, and have not even filled the role of principal opposition party in parliament. On the other hand, neither are politically negligible. The New

Democratic party has held government in its own right in the Canadian provinces of Manitoba, Saskatchewan, and British Columbia; and in the federal parliament during 1972–4 it held the balance of power, winning significant policy concessions from the government of the day. The Irish Labour party, for its part, has been described as being one of the 'three main parties', if a 'poor third';[5] and, on a number of occasions from the 1940s to the 1980s, has been the junior partner in coalition governments with Fine Gael (United Ireland party), the principal opponent of the larger Fianna Fáil (Soldiers of Destiny).

There is no uniformity in the organizational level at which trade unions affiliate to Labour parties. In the United Kingdom, Ireland, Canada, and New Zealand, they may affiliate both nationally and locally, and thus secure direct representation on major party organs at both levels. In Sweden and Norway, affiliation can occur only at the local level, and direct union representation is accordingly restricted to this level. Affiliation to the Australian Labor party, on the other hand, is possible only at the intermediate level of state branches, affiliated unions being directly represented at state conferences but not on either the federal or the local organs of the party.

No Labour party boasts the affiliation of all available trade unions. Indeed, in only one case, the Australian, does it appear that even a majority (about 60 per cent) of all available trade union members are collectively affiliated through their unions to a Labour party. The corresponding proportion is a little under half in both the United Kingdom and Ireland, but drops to about a quarter in Sweden and New Zealand, possibly 15 per cent in Norway, and a little less in Canada. On the other hand, *in all cases* collectively affiliated trade unionists outnumber the members of Labour parties enrolled as individuals. They account for well over 90 per cent of the Australian and Irish party memberships, a little under that for the British, and about 80 per cent for the New Zealand. The figure for the Swedish Social Democratic party is in the mid-seventies, and for the Canadian New Democratic party, in the mid-sixties; then there is a sharp drop to only a little over 50 per cent in the case of the Norwegian Labour party. As these proportions might suggest, the affiliation fees paid by trade unions represent a major source of party income in all cases.

[5] Chubb, *The Government and Politics of Ireland*, pp. 76, 83.

It has often been claimed, at least in the English-speaking cases, that the affiliation linkage entails trade union 'control' of the party. The preponderant weight of the unions' affiliated membership and of their financial contributions tends to be taken for granted as proof of the proposition. The British Labour party is the extreme case in that, owing to its constitutional arrangements and use of the card-vote system, there is a precise correspondence between the trade unions' affiliated membership (almost 90 per cent) and their voting strength at the party's annual conference. In addition, the affiliated trade unions are constitutionally guaranteed, in effect, a decisive majority (just on two-thirds) in its national executive body. These majorities, however, impressive and entrenched though they are, do not in practice translate into trade union control of the party, as distinct from a situation in which the party's parliamentarians and constituency branch memberships have an influential hand. And the main reason they do not, as Martin Harrison demonstrated at length some time ago,[6] is because the ideological spread of the affiliated trade unions means that they are typically incapable of acting *en bloc* in relation to issues that seriously divide the party. This is demonstrably true also of other Labour parties in which the numerical preponderance of affiliated trade unionists is less faithfully reflected in the composition of key party organs.[7]

The reverse charge, that the affiliation linkage entails party 'control' of the trade unions, is more rarely voiced but sometimes surfaces during factional battles, usually in the form of an out-group's claim that union leaders are unhealthily responsive to the wishes of the party. Invariably, however, the charge rests on considerations which are not intrinsic to the affiliation linkage as such. There is no question that Labour parties, especially when in government, have at times strongly influenced trade unions affiliated to them. On the other hand, there has been in this none of the continuing party control that is a routine feature of both party-ancillary and state-ancillary trade union movements in which, almost invariably, no formal affiliation is involved.

This point about the realities of party influence, and yet the absence of continuing party control, applies equally to those autonomous trade union movements where the affiliation linkage is

[6] Harrison, *Trade Unions and the Labour Party since 1945*, esp. ch. 5.
[7] See e.g. Martin, *Trade Unions in Australia*, pp. 115–18; Milne, *Political Parties in New Zealand*, pp. 113–14; Galenson, *Comparative Labor Movements*, p. 156.

unavailable. In all these cases the main trade union confederation has a special relationship with at least one major party. One confederation, in fact, has formalized that relationship in a way that would allow literal-minded constitutionalists to conclude that the party controlled the confederation. This is the Austrian OGB which, like the Confederation of Venezuelan Workers and the Israeli Histadrut, officially permits the participation of political parties in its governmental structure. All of the four significant Austrian parties have taken advantage of this. The allocation of positions on the OGB's central executive and on the executive committees of its sixteen member-unions is made, from party lists, by way of an electoral process which has consistently given the lion's share on all OGB bodies to the Social Democratic party—while, at the same time, conciliating minority political interests by effectively over-representing the People's (formerly Social Christian) party, the Communist party, and a non-partisan grouping. There is no doubt that the OGB's ties with the Social Democratic Party are very close. Major trade union leaders customarily occupy senior positions in the party's structure and hold parliamentary seats as party representatives. They have also held ministerial posts in its name. Nevertheless, the distinction between the OGB and the party has been carefully maintained by union leaders. They have been sensitive to the necessity of avoiding too intimate an identification with the party if the OGB's position, as a comprehensive and basically non-party confederation, is to be preserved. One reflection of this is the OGB's recognition of other parties in its electoral process; another is the existence of OGB officials who occupy parliamentary or other public positions as representatives of the People's party. Not only is it implausible to detect any element of continuing party control in the OGB's relationship with the Social Democratic party, but—beyond control—it is not even clear that the party has maintained the general ideological and organizational superiority which Klaus von Beyme described as always having characterized 'the Middle European type of interaction between the party and the unions'.[8] Jack Barbash, for one, has argued that since the war 'the OGB has replaced the party as the senior partner in the labour movement'.[9]

[8] von Beyme, *Challenge to Power*, p. 248.

[9] Barbash, 'Austrian Trade Unions and the Negotiation of National Economic Policy', p. 382.

The OGB's incorporation of parties has no exact counterpart in any of the other four autonomous movements in which the affiliation option is absent: Denmark, Finland, West Germany, and the United States. The Danish case, however, is distinctive in its own right because the main trade union confederation, the LO, is *constitutionally* linked to one party, the Social Democratic party. Two seats on the LO's executive body are specifically reserved for party representatives; and the party gives matching representation on its executive to the LO. Otherwise, the linkages depend primarily on overlapping memberships and informal consultations —but these, of course, like the long-standing Swedish convention that the LO president should be a member of the Social Democratic party's executive, can be just as compelling as formal requirements.

The Finnish case is distinctive for another reason. Finland has a party system that is highly fragmented by the standards of other countries with autonomous trade union movements. One outcome is the presence of two electorally significant parties on the left—the Social Democratic party and the Communist party (formally, the Democratic League of the People of Finland). Unlike its counterparts in other autonomous movements, the main trade union confederation, the SAK, is thus confronted with *two* major parties which can be regarded as more or less 'natural' allies. In the result, the trade unions at large, and the SAK in particular, have close links with both, especially through overlapping memberships. But the cost is a severely factionalized trade union movement which resembles a party-ancillary movement at the workplace level where, within the one trade union, there are normally two, or more, organizational sections divided in terms of party support.

In West Germany and the United States, on the other hand, the party systems are less fragmented than that of Finland, and there is nothing resembling the formal connection between the Danish LO and the Social Democratic party. German trade unionism, however, has a long tradition of party-political identification. The unions which the Nazis destroyed in 1933 were divided confederally along party-ancillary lines. It was in defiance of this tradition that the DGB, the dominant post-war confederation in West Germany, was expressly established on a non-party basis. Despite that, the DGB and its affiliated unions have moved into a close, if informal, relationship with the Social Democratic party. There is an overwhelming overlap between membership of the party and the officials of the DGB and its affiliated unions (something of the order of 90 per cent). The party is also the recipient of a great deal of organizational and financial support from the DGB unions, although direct cash contributions appear to be rare. Nevertheless, the DGB has continued to maintain an officially non-party position. Some Christian Democratic trade unionists, unconvinced, broke away from the DGB in 1955 to form their own confederation, the Christian Trade Union Federation, which openly identifies with the Christian Democratic–Christian Social Union (CDU–CSU). But other Christian Democrats did not join the breakaway, which has never been in a position to mount a really serious challenge to the DGB. The importance attached to retaining the allegiance of the DGB's substantial minority of Christian Democratic supporters is indicated by the convention that ensures the election of two Christian Democrats to most DGB and union executive boards, at all levels. DGB officials, as a result, are to be found not only in the councils of the Social Democratic party, but also in those of the CDU–CSU, both inside and outside parliament. Moreover, as the DGB has demonstrated on a number of occasions, there is no question of its slavishly following the Social Democratic party's lead. By the same token, the high proportion of party offices held by DGB union officials does not signify anything like union control of the party—'because, among other reasons, union officials never acted as a unified bloc within the party'.[10]

The political tradition of the American trade union movement is quite different from the German, as might be expected in 'the only industrial society that has not produced a working-class party'.[11] Historically, the general run of trade union leaders in the United States preferred to steer clear of close or consistent relationships with political parties. But, like their West German counterparts, they too have shifted decisively away from their tradition since the Second World War. By the mid-1950s most of them had moved into an association with the Democratic party which has not only proved durable but, if anything, has intensified. At the national level, there are linkages involving informal consultation and advice between party leaders and major trade union officials. The tip of this iceberg is the now customary official endorsement (virtually automatic—given a hiccup in 1972—since 1952) of the Democratic

[10] Willey, 'Trade Unions and Political Parties in the Federal Republic of Germany', p. 45.

[11] Sartori, *Parties and Party Systems*, p. 192.

presidential candidate by the AFL–CIO and many major trade unions. Trade union association with the Democrats is standard, too, at the state and local political levels. Variations from this pattern typically involve specific Republican candidates with a special appeal to labour. The Democratic party, like its Republican opponent, is not a mass party in the Social Democratic mould, and this can facilitate the influence of union leaders in party concerns, especially perhaps at the local level. Thus, whereas in Chicago the unions have customarily provided useful support to a dominant party machine, in Detroit the unions have virtually *been* the machine; and elsewhere, in the absence of party branches, union locals have often acted in the Democratic cause. Again, whereas in many cases trade union leaders simply extend support to such local, state or federal candidates as the party may select, in others they have a hand in the choice through consultative muscle or membership of formal selection bodies; and in still others, they may determine the selections. Indeed, it has been claimed precisely that 'organized labor determined the party's presidential candidate' in 1984.[12]

It is true that some American trade union leaders are Republican supporters. It is also true that many still adhere to the traditional policy of avoiding consistent commitment to either party. But the prevailing tendency is reflected in the existence and activities of the Committee on Political Education (COPE). COPE, an offshoot of the AFL–CIO, presides over a network of state and local COPEs; and is concerned with public elections at all levels. Its principal tasks are to raise election campaign funds and to organize the labour vote in support of favoured candidates, usually Democratic party nominees. It, and its outriders, solicit money from union bodies and co-ordinate its distribution. The emphasis, however, is on the provision of organizational services (manpower, mailing, telephones, computerized voter data, and other facilities) to ensure, first, that union members and other selected groupings register as voters; and second, that they actually turn out to vote. In this way the AFL–CIO, through COPE, has furnished the Democratic party with the 'one thing [that] business interests have been unable to supply for the Republicans, namely, organization . . . the sheer muscle union workers can provide in a campaign'.[13] Union

[12] Miller, 'The Mid-Life Crisis of the American Labor Movement', p. 168.

[13] Masters, 'The Organized Labor Bureaucracy as a Base of Support for the Democratic Party', p. 258.

organizational help seems to have been of particular importance in national elections (though there are also many parts of the United States where candidates are reputed to prefer COPE's money to its organization because of unfavourable public attitudes towards trade unionism). All these developments have been interpreted as essentially involving 'labor's organizational integration into the Democratic party', but without entailing that either the party or the unions dominate the other; instead, they operate in 'a political coalition as partners'.[14]

RELATIONS WITH THE STATE

There are two fairly distinct dimensions to the relationship between autonomous trade union movements and the state. One has to do with the state's formal attempts to regulate trade union activities, as reflected in the law relating specifically to trade unions. The second has to do with dealings between the state and the trade unions which (while they may be regulated or structured by the law in some degree) are political rather than legal in character, in that they centre on consultation and negotiation.

Both these dimensions have increasingly come to be focused on issues at the heart of trade union concerns. The central official preoccupation of the trade unions comprising autonomous movements is the pay and working conditions of their members. This is reflected in the weight they place on collective bargaining with employers (private and public) as the process through which they have a hand in the regulation of these issues.

The Law

Law, of course, may be promotional or protective of an interest; and there is a significant element of that in the laws under which most autonomous trade union movements operate. The other side of a legal relationship with the state is the extent to which the law purports to regulate or control an interest. For trade unions in autonomous movements, what counts above all is their freedom (within limits set by their tactical opportunities and 'natural' strategic position) to share in the regulation of their members' pay and working conditions by way of collective bargaining with employers.

[14] Greenstone, *Labor in American Politics*, pp. 10, 318.

There are three principal forms which legal restrictions on this freedom can take. One involves government intervention in the process of collective bargaining itself. Another involves limitations on the unions' use of the strike as a bargaining weapon. The third involves government intervention in the unions' internal affairs (with the ultimate implication of a voice in their bargaining decisions).

All autonomous trade union movements are subject to legal regulation in one or more of these forms. But only four, all in English-speaking countries (United States, Britain, Australia, New Zealand), are subject to all three forms of regulation in a substantial sense—and, in particular, to the least common form involving significant intervention in unions' internal affairs. Thus, there is legislation in each of these four countries concerned with the selection of union officials (specifying such things as necessity and periodicity of elections, electoral procedures and/or providing means for dealing with alleged election irregularities). In the United States, unionists with a particular political tie (membership of the Communist party) are debarred from holding union office. The raising and expenditure of funds for political purposes is statutorily regulated (United States, Britain, New Zealand), as is the manner in which union dues may be increased (United States). Union rules are open to judicial examination and disallowance on the ground of being 'unreasonable' or 'oppressive' to members (Australia, New Zealand); and members are statutorily guaranteed specific rights in relation to participation in union affairs (United States).

As for intervention in collective bargaining, the general tendency in countries with autonomous trade union movements has been for the state to leave the field, legally speaking, to the voluntary action of unions and employers. In Britain this extends to the point that collective agreements are not enforceable at law—a position which applies in Ireland as well, although there is a procedure, not often used, for converting agreements there into legal instruments. The most pronounced deviations from the voluntarist tendency occur in Australia and New Zealand where the state formally intervenes in the collective bargaining process by way of comprehensive systems of compulsory arbitration presided over by state-appointed tribunals. Compulsory arbitration has also been used, but in a very much more partial and limited sense, in Norway and in Canada. In most other cases, government conciliation or mediation facilities are available, being least noticeable in West Germany. Use of such services may be compulsory during the initial period of a dispute before a strike can take place (Canada), and the same result may be achieved by the legal power of a mediator to prohibit strike action for a limited period (Norway, Denmark, Finland). In all other cases, resort to state arbitration or conciliation services is generally voluntary.

In all but one of the English-speaking countries (Britain), the law's involvement in the collective bargaining process goes as far as conferring monopoly rights on recognized trade unions where the recruitment and representation of specific groupings of employees is concerned. In three of these cases, the monopoly grant turns on the decision of either a state tribunal (Australia, New Zealand) or a minister (Ireland). In two, it depends upon a ballot of the workforce affected (United States, Canada). There is a similar element of monopoly in the procedure available in three other cases (Austria, West Germany, Denmark) where, by government decision, the legal scope of a collective agreement may be extended to cover employers and employees beyond those technically party to its negotiation. But otherwise, government involvement in this kind of rule-making, to do with the *procedures* of collective bargaining, is much more pronounced in the English-speaking countries (Britain excepted) than in the Continental.[15]

All autonomous trade union movements are subject to legal restrictions applying to strike action under either statute (federal or state, or both, in the case of federal systems) or judicial decision—the latter totally dominating the field in the West German case. Particular types of strike are outlawed in different cases, such as unofficial strikes (West Germany, Sweden), political strikes (West Germany), secondary boycotts (United States, Britain, Australia), certain public employee strikes (Austria, West Germany, Norway, Britain, Canada, United States), strikes against the employment of women or to enforce a closed shop (Ireland), and strikes during the lifetime of an agreement on an issue already determined by it (Canada, Sweden, Norway, Finland, West Germany, Austria). Otherwise, strikes must first be authorized by a ballot of the employees concerned (Britain, West Germany, Australia), advance notice of them must be given a specified time beforehand (Sweden,

[15] See e.g. ILO, *Collective Bargaining in Industrialised Market Economies*, p. 139.

Norway, Finland, Australia), or they must be preceded by mediation proceedings (Canada). Legislation relating to strikes in 'essential industries', or otherwise of an 'emergency' character, enables government to impose a 'cooling-off' period involving a return to work (United States, Canada, Norway, Finland), or to take other action against such strikes (Britain, Australia, New Zealand). The enactment of special legislation to deal with specific 'emergency' strikes is not uncommon in at least two cases (Canada, Denmark).

All of this adds up to a picture, in the broad, of complex and detailed legal regulation affecting each of the twelve main autonomous trade union movements. But, then, there are the variations in this respect. And of the twelve, one movement—the Australian—is subject to a pattern of regulation which is more comprehensive and, on the face of it, more restrictive than in any of the other cases. Australia, in other words, represents the extreme in terms of formal state intervention; and it was in partial recognition of this that Selig Perlman, a half-century ago, drew a striking (if questionable) comparison with his remark that 'Australian labor is the best organized labor outside Soviet Russia'.[16] The Australian case is thus the limiting case among autonomous trade union movements.

The legal controls on Australian trade unions stem from legislative adoption of the principle that compulsory arbitration is preferable to free collective bargaining as a means of resolving serious industrial disputes. The outcome has been formal government intervention, on a massive scale for almost the whole of this century, in the collective bargaining process. A complicated mosaic of tribunals and courts, appointed under either federal or state legislation, administer a system (technically, seven systems) of compulsory arbitration which requires them to formulate, interpret, and enforce a host of awards (and to approve industrial agreements, otherwise not legally enforceable) dealing with most aspects of the employment relationship, often in elaborate detail. The pay and working conditions of over 85 per cent of all Australian employees are directly affected in some measure by these awards and approved agreements.

Registration is a pre-condition of access to all the major tribunals, so far as individual trade unions are concerned. Virtually every Australian trade union is registered, in one or more

[16] Perlman, 'The Principle of Collective Bargaining', p. 157.

jurisdictions. Simultaneous state and federal registration is not unusual; and different members of the one union may be covered by either a state or a federal award. Almost all unions of industrial or numerical significance are registered at least under the main federal tribunal, and are thus subject to the federal legislation. On the other hand, owing to the intricacies of the Australian Constitution, their state branches may altogether escape being so.

The federal arbitration system occupies a position of special influence, even though it directly covers little more than one-third of all employees (as against the one-half that come under state awards and agreements). On the whole, the federal system has tended to act as pace-setter for the state systems—in relation to both statutory provisions and, more particularly, arbitration decisions. The main federal tribunal, moreover, occupies a unique position as a result of the terms of the Australian Constitution. For its decisions, and the industrial matters with which they are concerned, are beyond the direct legislative reach of the federal government. This extraordinary independence is not shared by state tribunals: there is no similar constitutional bar preventing state governments from legislating directly on industrial matters, if they wish.

For over eighty years, the main federal tribunal has officially held to the proposition that strikes are incompatible with the spirit of the arbitration system, although the legislation under which it operates has not expressly prohibited strikes since 1930. On the other hand, the powers conferred on the tribunal do enable it to prohibit and to penalize strikes in specific cases, while other federal legislation provides sanctions against strikes involving secondary boycotts. Under both federal and state legislation, sanctions are also available in relation to strikes in a wide range of defined 'essential' industries. In addition, most state arbitration statutes either prohibit strikes outright or specify that they are legal only if a specified condition has been met (alternatively, a 'cooling off' period or a secret ballot). In one state, strikes on certain issues (demarcation, union security, technological change) are altogether illegal. The penalties available under this array of anti-strike provisions include, in the case of trade unions as organizations, fines and cancellation of registration; and, in the case of individual officials and members, imprisonment, fines, and the denial (e.g., of a pay rise granted other groupings) or deprivation (e.g., of long-service leave entitlements) of advantages. In addition, Australian trade unions are generally vulnerable to civil actions for damages arising from a strike.

Legislative intervention in the internal management of Australian trade unions is most far-reaching in the federal jurisdiction, which provides a model that is followed in varying degree by corresponding state legislation. The rules of federally registered unions are required to deal in specified ways with certain matters, including admissions to membership, resignations and, above all, the methods of electing union officials—secret postal ballots are to be used for all rank-and-file elections; full-time officials must be selected either by a rank-and-file ballot or by a smaller body whose members are themselves elected by the rank and file; and no official, full-time or part-time, may hold office for more than four years without standing for re-election. The leadership of a registered union may entrust the conduct of a postal ballot to a government official (in which case the substantial cost involved is met from public funds), and the same result may be achieved by a specified number of ordinary members opposed to a union-appointed returning officer. Members may also secure a judicial inquiry into the conduct of an election; and, in the event of substantial irregularities, the result may be changed or a new election ordered. More generally, there are also procedures enabling members to have a union rule subjected to judicial scrutiny. In such cases, if considered 'unreasonable' or 'oppressive' to members, the rule may be altered or cancelled by the court; or, if the court interprets the rule differently from the union's leaders, they may be required to comply with its interpretation.

The formal picture of the statutory context within which Australian trade unions operate is thus one of a highly intrusive legal framework. It is, however, a picture that needs to be qualified in two ways. In the first place, there are aspects of the law which, from the trade union viewpoint, are of positive value—specifically, the legal and other advantages which can be shown to flow from registration.[17] This helps balance the picture by making it clear that the legal framework is not simply a matter of control and regulation, but involves a substantial component of protection and promotion as well. In the second place, and for present purposes

[17] See e.g. Martin, *Trade Unions in Australia*, pp. 38, 110–11.

more significantly, the pattern of legal regulation is in fact nowhere near as formidable in the spirit as it appears to be in the letter.

Take the case of compulsory arbitration. As the system operates in Australia, it leaves a great deal of room for what amounts to free collective bargaining between unions and employers. For a start there is, in both federal and state jurisdictions, substantial emphasis on conciliation as distinct from arbitration, reflected particularly in a disposition on the part of formal arbitrators to strive for agreement between the parties before resorting to an arbitral role. Then there is the frequency with which arbitrators have accepted that conditions established earlier or elsewhere by way of collective bargaining should be incorporated in their awards—or, at least, ought to be taken into account in the course of arriving at related decisions. But above all, there is the readiness that arbitrators have shown to convert privately negotiated agreements into awards (informally known as 'consent' awards). The precise incidence of such awards is not known, but it would be no surprise if it were to match the 94.5 per cent of all awards which, it has been calculated, were consent awards between 1975 and 1981 in New Zealand.[18] In any case, the Australian arbitrator usually decides only one or two disputed issues in the case of awards which are actually, as well as technically, the outcome of an arbitral decision—most of their clauses being agreed to by the unions and employers concerned.

The gap between the letter and the application, however, is most glaring in the case of strike law. One thing is quite clear: throughout this century, in which compulsory arbitration has technically reigned supreme, strikes have become an increasingly common feature of the Australian industrial scene.[19] In addition, although international comparisons are hazardous owing to national variations in the basis on which strike statistics are collected, it seems clear that strike activity in Australia is not unusually lower (nor unusually higher) than that in other politically and culturally similar countries that lack compulsory arbitration systems.[20] Yet, overwhelmingly, the Australian strikers and trade unions involved have been untouched by legal sanctions, although usually technically vulnerable to them. Serious attempts to enforce the anti-strike

[18] See Geare, *The System of Industrial Relations in New Zealand*, p. 58.

[19] See Waters, *Strikes in Australia*, pp. 49–52.

[20] See Beggs and Chapman, 'Australian Strike Activity in an International Context', pp. 146–8.

provisions of the law have been relatively few and, for the most part, sporadic because employers, governments, and industrial tribunals have tended, in general, to be wary of using them. This gap between the letter and the application of the law is not a peculiarly Australian phenomenon. Much the same can be said in the case of New Zealand, with its compulsory arbitration system; and the enforcement of less comprehensive anti-strike sanctions has been similarly deficient in, for example, the United States, Austria, and West Germany.[21]

It is a different story, however, in the case of legal controls over the internal affairs of Australian trade unions. The federal provisions, in particular, have been applied with notable frequency. Although usually enacted for partisan political reasons (above all, in the purported belief that they would help dislodge or contain Communist union leaders), they have plainly enhanced union democracy in the sense that the actions and positions of all trade union leaders, irrespective of political colour, are more open to challenge from within their membership than would otherwise be the case. This essentially democratic element is underlined, in the federal jurisdiction, by the provision of government financial support for the legal costs of a union member with a prima-facie case against the rules of his union or the conduct of its officials. There is no doubt that it is the ability of out-factions to embarrass trade union leaderships, rather than state control, which has been fostered by legislative provisions such as those enabling inquiries into union-conducted elections. The proof of the pudding lies in the fact that the federal legislation on elections and union rules, while once hotly opposed—especially by those on the left wing of the trade union movement—has since come to be used in factional battle as much by the left as by the right. Its provisions are no longer controversial in trade union circles. The critical point, in any case (so far as the issue of state domination is concerned), is that legal intervention in the internal affairs of Australian trade unions stops well short of allowing routine government interference in either the selection of their officials or the management of other domestic matters.

The Australian case, as already pointed out, represents the

[21] See Geare, *The System of Industrial Relations in New Zealand*, p. 299; Blanpain, *Comparative Labour Law and Industrial Relations*, p. 512; and Tomandl and Fuerboeck, *Social Partnership*, pp. 55–6.

extreme among autonomous trade union movements so far as formal state intervention is concerned. Australian trade unions, legally speaking, are confined within a legislative strait-jacket when it comes to collective bargaining, to striking, and to key aspects of their internal management. As a matter of practice, however, the legal provisions involved are either imperfectly enforced or, for the greater part, do not impinge on the unions' independence of action in a way, or to an extent, that seriously troubles most of their members, their leaders, or the factional groupings that struggle for office within them. And that illustrates a larger point about the kind of societies in which autonomous movements exist—they are societies in which the legal regulation of group activities, whether they be trade unions or other corporate bodies, generally tends to be much harsher in the letter than in the application.

Working Relations

All autonomous trade union movements are engaged in continual consultations and negotiations with government and its agencies. Nowadays, too, the main or sole national confederation usually plays the major part in this. For the trade unions in general, their interest in government policy and administration tends to take second place to their interest in collective bargaining about pay and working conditions—except when (as in the formulation and enforcement of incomes policies) government action is directly concerned with pay and working conditions. On the other hand, government policy and administration are invariably of central concern to the major confederations in autonomous movements. Their leaders, as a result, characteristically display a keen interest in dealing with existing and, often, potential office-holders in the institutions of government. They are usually prepared to deal with governments whatever their political colour.

The form and intimacy of union–government dealings, and even the policy areas focused on, vary from country to country. The variations are probably even greater when it comes to the influence which trade unions are able to exert on government policy and administration in the different national cases (although comparisons in this respect are necessarily tentative). Even within the one national case, there tend to be variations over time in the trade unions' influence and (to a lesser extent) in their consultative arrangements with government. These variations turn especially, but not only, on the colour of the governing party.

In the United States the trade unions' relations with the federal government are distinguished by an emphasis on direct dealings with individual members of the legislature (Congress), at the level of both the Senate and the House of Representatives. The character of the American presidential system, as determined by a constitution which entrenches an extreme version of the separation of powers principle, makes for a legislature that is far more independent of the executive, generally speaking, than in the case of the parliamentary systems operating in all the other countries with autonomous trade union movements. In parliamentary systems, the political leadership of the executive usually exercises greater control over the legislature. This is reflected in tighter party discipline and, correspondingly, the diminished political importance of individual members of the legislature. In these systems, as a result, it is the executive (ministers and bureaucrats) which trade union leaders tend to focus on in their dealings with government. Exceptions to this rule, however, are by no means uncommon. By the same token, American trade union leaders have not concentrated on Congress to the total exclusion of the executive—as indicated by the numerous personal meetings and telephone conversations between President Johnson and George Meany, the president of the AFL–CIO.[22]

The channels of communication between trade union leaders and government office-holders (legislative or executive) in all countries range from informal, *ad hoc* dealings, such as casual meetings and telephone calls, to formal consultative procedures involving official advisory and administrative bodies. Informal dealings probably receive most emphasis, and consultation through formal bodies least emphasis, in the United States. Contact with members of Congress there is maintained mainly through professional lobbyists acting for the unions, the AFL–CIO employing seven full time in Washington and most major national unions retaining at least one or two. Formal consultative bodies tend to play a greater role in the parliamentary systems, but their significance and number vary from country to country and (generally as a result of changes in government) over time. In their advisory form they appear to be most prolific, comprehensive, and entrenched in the case of Austria.[23] Consultative bodies are usually tripartite in that they

[22] von Beyme, *Challenge to Power*, p. 235.

[23] See Katzenstein, *Corporatism and Change*, pp. 76–7.

include representatives of employers' organizations, as well as of trade unions and government. Government representation on them is quite often at the ministerial level.

The issues raised in dealings between trade unions and government office-holders can range, from time to time, over most major aspects of domestic policy and quite commonly into foreign policy. But invariably they cluster in the economic and social welfare areas. Traditionally, they tended to exclude the central concerns of the unions, in the sense that the detailed regulation of pay and working conditions was left largely to collective bargaining—or, as in Australia and New Zealand, to a mixture of collective bargaining and compulsory arbitration. Not that governments refrained altogether from direct intervention in this area. On the contrary, in different cases there is legislation—much of it long-standing—specifying such things as minimum wages (United States, Canada, New Zealand), maximum working hours (Sweden, West Germany, Austria, Ireland, United States, Canada, Australia), dismissal procedures (West Germany, Austria, Sweden, Britain, Ireland, Canada, Australia), paid holidays and vacation entitlements (Austria, Sweden, Finland, Ireland, Canada, Australia, New Zealand). But, since the Second World War, the traditional exclusion of pay and working conditions from union–government dealings has been breached in a much more fundamental respect as a result of the interest which liberal democratic governments have displayed in wage-restraint as an instrument of national economic policy. The development of incomes policies—and, above all, trade union involvement in their application—has been construed as the primary symptom of a neo-corporatism which, arguably, has at least budded, and occasionally flowered, within liberal democratic political systems in the second half of the twentieth century.[24]

The theory of neo-corporatism, for all its confusions,[25] perceptively fastens on one striking, but not consistent, feature of government–trade union relations in recent times. And that is the willingness, or otherwise, of trade union leaderships, in peace time, to enter *voluntarily* into arrangements with government which involve limiting the wage rates that their members (or at least some of their members) could otherwise expect to secure. It is the voluntary element which is the key indicator of the degree of corporatism.

[24] See esp. Cawson, *Corporatism and Political Theory*.

[25] See Almond, 'Corporatism, Pluralism, and Professional Memory', pp. 251–60; Martin, 'Pluralism and the New Corporatism', pp. 86–102.

Thus corporatism flourishes where (and when) incomes policies are formulated and, above all, enforced with the willing co-operation of trade union leaderships. Alternatively, corporatism is weak or non-existent to the extent that incomes policies have to be imposed unilaterally by government, and depend upon legislative compulsion. On this yardstick, although transnational rankings must admittedly be tentative,[26] a rough ranking of countries with autonomous trade union movements is possible. Corporatism has been decidedly weak in the United States, Canada, and Denmark, where incomes policies have been almost wholly a matter of compulsion, and relatively brief in their duration. New Zealand constitutes a somewhat more durable version of the same category. The opposite extreme, involving long-term incomes policies formulated and applied on a consensual basis, is represented above all by Austria (with its proclaimed 'social partnership')[27] and, less consistently, by Sweden and Norway. The medial cases range from mainly consensual arrangements (West Germany, Finland, Ireland) to a more even mix of the consensual and the compulsory (Britain, Australia).[28]

Voluntary participation in an incomes policy entails an explicitly administrative role for unions, in so far as they are concerned with enforcing the terms of the policy by ensuring the compliance of union members. Autonomous trade union movements may be incorporated in the administrative structure of government in other respects as well: a notable and extreme example is the function discharged by Swedish and Danish trade unions (and once handled also by British unions) in administering the state's unemployment benefits system.

One final point, arising from the peculiar sanction possessed by trade unions. Dealings between autonomous trade union movements and government may involve more than negotiation and consultation. They may involve strike action against government or about government policy—and not only by government employees concerned with their conditions of employment, but in the shape of

[26] See e.g. Lehmbruch and Schmitter, *Patterns of Corporatist Policy-Making*, pp. 8–12.

[27] See Tomandl and Fuerboeck, *Social Partnership*, esp. pp. 23–6.

[28] This ranking is derived mainly from Marks, 'Neocorporatism and Incomes Policy in Western Europe and North America', pp. 270–3, apart from Australia and New Zealand which are not included in his data. See also, in this respect, Lehmbruch and Schmitter, *Patterns of Corporatist Policy-Making*, pp. 16–21, 245, 257.

'political' strikes where government is not playing the role of employer.[29] Political strikes are not uncommon among autonomous movements in general, though perhaps least so in the American case. But, since the early 1970s at least, it appears that they may well have been employed more widely and more frequently in Australia than in any other case.[30] This provides a further indication of the independence of Australian trade unions despite the exceptionally comprehensive legal controls to which they are formally subject.

SUMMATION: VARIETY AND INDEPENDENCE

Variety is a characteristic of autonomous trade union movements. They exhibit a remarkable diversity in terms of structure, functions, and formal relations with both political parties and the state. In the end, however, their common identification involves an essential uniformity. They are all, in comparative terms, effectively independent of parties and governments; and equally, it must be stressed (given the surrogate position as a possibility), parties and governments are effectively independent of them. This is not to deny that in both relationships there are at least occasional, and often fairly consistent elements of dependence. The point is that these elements do not add up to a pattern of dominance or control in anything more than, at the most, a temporary or selective (as to issue or area) sense.

In their relations with parties, as we have seen, the solo or main confederations in autonomous trade union movements characteristically maintain closer relations with one major party than with others. On the other hand, they are typically eager both to avoid too intimate an indentification, and to assert a basic political neutrality when it comes to dealing with matters impinging on their members' industrial concerns.

In their relations with governments, such confederations characteristically deal with government office-holders (if often neither from a position of strength nor, in their own terms, successfully) at least with the status of one of the more prominent interest groups of independent means. The formal government incomes policies with

[29] On the concept of a political strike, see Martin, 'The Problem of "Political" Strikes', pp. 68–81; Aaron and Wedderburn, *Industrial Conflict*, ch. 6.

[30] See Martin, 'Political Strikes and Public Attitudes in Australia', pp. 269–81.

which autonomous trade union movements have been associated, whether voluntarily or compulsorily, implicitly acknowledge their essential independence. This acknowledgement takes one form in the case of consensual policies, which normally involve the negotiation of trade-offs in exchange for union co-operation in restraining wages. It takes another form in the case of unilaterally-imposed policies: in countries with autonomous trade union movements, such policies have tended to be much less durable than their consensual counterparts.

14

SOME EXPLANATIONS

The preceding chapters have sought to establish a workable classification of national trade union movements. To this end, they delineated the leading characteristics of each of five categories, and illustrated them with particular reference to 27 specific cases.

The present chapter takes the next step. It is concerned with the theorizing aspect of the study. That is to say, it seeks to identify the principal factors moulding the five different types of trade union movement. Pre-eminent among these causal factors appears to be the nature of the political system that a trade union movement inhabits.

PARTY SYSTEMS

Classifications of modern political systems have tended to focus on political parties. Perhaps the most sensible and usable typology of party systems is that developed by Giovanni Sartori in his brilliant *Parties and Party Systems*.[1] What follows leans heavily, but not without qualification, on his work.

At the most general level, there is a distinction to be drawn between political systems in which party organization is a factor in the exercise of state power, and systems in which it is not. Systems of the latter kind, 'no-party' states, take two forms: party-less and anti-party. Party-less states are usually associated with traditional societies (such as Saudi Arabia): but most anti-party states, according to Sartori, are associated with military regimes in developing countries[2]—though this is not to say that military regimes, as such, are anti-party.

Political systems reliant on parties generate a more complicated taxonomy, with two primary levels. The first level involves a

[1] Other typologies (with some of which Sartori's overlaps) range from the over-simple to the over-refined, involve more or less serious problems in application and, in some cases, produce distinctly odd bedfellows.

[2] Sartori, *Parties and Party Systems*, p. 40.

distinction between states with *competitive* party systems and those with *non-competitive* party systems. 'Competition', Sartori defined as existing when elections are contested, or may be contested, under circumstances in which all candidates are able to nominate 'without fear and with "equal rights" '.[3] At the second level, competitive party systems are divisible into four principal categories; and non-competitive systems into two.

Competitive

Sartori's titles for the four competitive categories are *predominant-party* system, *two-party* system, *limited (or moderate) pluralism,* and *extreme (or polarized) pluralism.* It is to be emphasized that, in the case of all these categories, the raw number of parties, though significant, is not in itself the critical criterion. In particular, there are qualifications arising from two considerations. One has to do with the 'relevance' of different parties—a notion explained below in connection with the category of limited (or moderate) pluralism. The other turns on the proposition that each type of party system is to be defined not merely in terms of the number of parties involved (its 'format'), but also with an eye to a 'set of properties' identified with it (its 'mechanics').[4] For example, a critical property of a two-party system is that the participating parties alternate in office.

1. *Predominant-party system.* This system is 'at the edge of the competitive area'.[5] It is characterized by a total or near-total absence of rotation in government office, as between parties, the predominant party continually winning control of government in elections. Crucial requirements are that the opposition parties are 'truly independent', and that their electoral failures 'cannot reasonably be imputed to conspicuous unfair play or ballot stuffing'.[6] There is, in other words, a genuine competitive element.

2. *Two-party system.* 'Two parties compete for an absolute majority that is within the reach of either'; and the winner controls government on its own, 'but not indefinitely', its replacement by the other remaining at least 'a credible expectation'.[7]

3. *Limited (or moderate) Pluralism.* There are at least *three* parties that are 'relevant' in the sense that they possess either 'coalition

[3] Sartori, *Parties and Party Systems*, p. 217. [4] Ibid. 188–9. [5] Ibid. 200.
[6] Ibid. 195. [7] Ibid. 127, 186, 188.

potential' (a tactical and ideological position that at some time enables them to share in a coalition government) or 'blackmail potential' (a 'power of intimidation') arising from an opposition party's electoral and parliamentary weight, enabling it to affect the behaviour of 'governing-oriented' parties.[8] Further, no party has a prospect of securing an absolute majority, and governments are *always* coalitions.[9] As compared with extreme pluralism, limited pluralism 'lacks relevant and/or sizable anti-system parties' (see below).[10]

4. *Extreme (or polarized) pluralism.* There are at least *five* relevant parties. As with limited pluralism, coalition governments are the rule. On the other hand, extreme pluralism is characterized, in particular, by relevant 'anti-system parties' (especially of the Communist or Fascist brands); and by 'bilateral oppositions', or 'counter-oppositions', consisting of two opposition party groupings which cannot join forces, and therefore oppose each other as well as the governing parties.[11] This type of party system is also distinguished by 'the centre placement of one party or of a group of parties', which forms the fulcrum of all possible government majorities, with the result that coalitions are normally confined to 'the centre-left and/or the centre-right parties only'.[12]

Non-competitive

Sartori's titles for the two types of non-competitive systems are *one-party* and *hegemonic-party*. Each, as he defined them, involves a single, permanently ruling party; but beyond that, these categories, and his application of them, raise serious problems. His primary definitions are as follows.

1. *One-party system.* 'Political power is monopolized by one party only, in the precise sense that *no other party is permitted to exist.*'[13]

2. *Hegemonic-party system:*

> The hegemonic party *neither allows for a formal nor a de facto competition* for power. Other parties are permitted to exist, but as second-class, licensed parties; for *they are not permitted to compete with the hegemonic party in antagonistic terms* and on an equal basis.[14]

[8] Ibid. 122–3. [9] Ibid. 127, 178. [10] Ibid. 179.
[11] Ibid. 132–4. [12] Ibid. 134, 139.
[13] Ibid. 221. Emphasis added. [14] Ibid. 230. Emphasis added.

The difficulty involved in each of these definitions is indicated by Sartori's application of them. The key cases, which he discussed at some length, are East Germany, Poland, and Mexico.

The problem of the one-party system definition. This, as quoted above, specifies that no party other than the ruling party is 'permitted to exist'. One of the countries Sartori designated as a one-party state was East Germany; one he excluded from that category (describing it as hegemonic-party) was Poland. Yet in both countries, as he acknowledged, there are a number of 'second-class, licensed' parties (as the *hegemonic*-party definition describes them)—four in East Germany and five in Poland. He distinguished the East German case (as a one-party system) from the Polish case on the ground that the German minor parties, unlike the Polish, were 'a pure sham, an empty facade'.[15] This description presumably applied as well (though Sartori neither specified this nor mentioned them) to the minor parties found also in three of the other Communist states—China, Czechoslovakia, and Bulgaria—which he expressly located in the one-party category.[16] Astonishingly, in these circumstances, the only evidence he provided in support of the Polish exception was a simple assertion by a Polish author that the five minor parties 'share governmental and administrative posts at all levels . . . [and] shape public opinion . . . but without attempting to undermine the position of the [ruling] party'.[17] And this is Sartori's only justification for concluding that the Polish parties are not 'sham' parties because the ruling party has allocated 'a fraction of its power' to them.[18]

There are two particular problems about Sartori's treatment of the one-party category. One is the sheer empirical uncertainty of his distinction between minor parties which are 'pure sham, an empty facade', and those which are not. The issue is important because 'second-class, licensed' parties are not at all unusual among non-competitive party systems at large. There are many states, technically multi-party, but with a single party that is overwhelmingly dominant and permanently monopolizes all key government positions. In such cases, it is usually not merely a matter of selected parties being *permitted* to exist, but of their being *promoted* through subsidy, services, and patronage by the ruling party.

[15] Sartori, *Parties and Party Systems*, p. 230.
[16] Ibid. 221.
[17] Ibid. 231.
[18] Ibid.

Among Communist countries alone, not only China, Czechoslovakia, and Bulgaria, as we have seen, but also North Korea and Vietnam each have one or more officially recognized minor parties which, like those in East Germany, 'acknowledge the leading role' of the Communist party 'and actively support the construction of socialism'.[19] If that kind of cosy, non-competitive relationship makes minor parties 'a pure sham, an empty facade', then (on Sartori's own showing) it is difficult to see how the Polish parties can escape the label—given that in Poland, too, as he put it, 'the very premises of competition are ruled out'.[20] At any rate, his discussion of these cases provides no practicable guidance as to what it is that distinguishes the sham from the non-sham in this respect.

The other problem about Sartori's treatment of the one-party category is more fundamental. As we have seen, he defined one-party systems 'in the precise sense' of a situation in which one party, and one party alone, is 'permitted to exist'. But then, as we have also seen, he admits to the one-party category political systems (East Germany, China, and the rest) in which there *are* other minor parties. And this is the problem. These parties may or may not be thought of as 'sham' and 'facade', but one thing about them is incontrovertible: they are 'permitted to exist'. Given that, they stand in flagrant contradiction of Sartori's definition of one-party systems.

The problem of the hegemonic-party system definition. This, as quoted above, specifies that 'second-class, licensed parties,' while allowed to exist, are not permitted 'to compete . . . in antagonistic terms' with the ruling party, even in a purely 'formal' way. Sartori designated both Poland and Mexico as hegemonic-party states. Poland, for its part, sits snugly under this definition—but, then, so too do the Communist states (East Germany, China, Czechoslovakia, and Bulgaria) with 'sham' parties, which he classified as one-party systems. For in all of them, as in Poland, the minor parties secure seats in the legislature by a process which avoids even the pretence of public competition: they are given representation on a common list, shared with the ruling party, in general elections where voters have the option of voting for or against the list.

[19] *GDR: 300 Questions 300 Answers*, pp. 30–1; and see Holmes, *Politics in the Communist World*, pp. 208–16.
[20] Sartori, *Parties and Party Systems*, p. 231.

Mexico, on the other hand, does not conform to the requirements of Sartori's hegemonic-party definition. It is one of a number of non-Communist states (such as Egypt and Singapore) in which formal competition *is* permitted by a permanent ruling party. Second-class parties are allowed to nominate separate candidates in elections in which voters can choose between candidates; or, as in Taiwan (before the formal admission of opposition parties to the electoral process in 1986), individuals could be nominated as 'non-partisan candidates'.[21] In each case, it is to be emphasized, the competition permitted is strictly limited and tends to attract legal and/or extra-legal action on the part of the authorities if it exceeds the bounds of the permissible, as the regime chooses to interpret it. For example, as in Mexico, opposition parties may be permitted to defeat the ruling party's candidates in elections so long as only minor positions are involved, and so long as the scale of their electoral successes is not seen to represent a threat to the ruling party. But the crucial point for present purposes is that second-class parties can, and do, formally stand candidates against ('compete . . . in antagonistic terms' with) the ruling party, and voters can choose between them.

It is true that, at one point, Sartori implicitly creates a niche within the hegemonic-party category which accommodates the Mexican case. This emerges in his remark that 'the hegemonic party formula *may afford the appearance* but surely does not afford the substance of competitive politics'.[22] Yet that remark flies in the face of his definition with its careful and measured exclusion of apparent ('formal') competition.

The distinction between one-party and hegemonic-party systems, according to Sartori's original definitions, is a distinction between systems in which second-class parties do not exist at all and, on the other hand, systems in which they do exist, while in no sense ('neither . . . formal nor . . . de facto') competing with the ruling party. But then, in each case, comes his radical and puzzling amendment. First, 'sham' second-class parties do not count as parties—with the result that the one-party category, which originally excluded both Poland and East Germany, now includes East Germany. Second, the 'appearance' of competition does not count

[21] Jacobs, 'Political Opposition and Taiwan's Political Future', p. 21.
[22] Sartori, *Parties and Party Systems*, p. 231. Emphasis added.

as 'formal' competition—with the result that the hegemonic-party category, which originally included Poland and not Mexico, now encloses Mexico as well.

These definitional confusions obviously need to be resolved if the two categories are to fulfil a useful function. This can be done by fastening on one quite unambiguous distinction evident in the case of the three countries discussed. The distinction lies in the fact that, while all three countries boast second-class parties, only in Mexico are the minor parties permitted to contest public elections against the ruling party. This fact sharply and unequivocally distinguishes the Mexican case from that of both Poland and East Germany. Moreover, the distinction turns on a consideration (the formal ability to contest elections) which is not only conceptually clearer than Sartori's sham-party distinction between Poland and East Germany, but is also infinitely simpler to observe and to substantiate. It is, as well, a consideration centred on something that plainly *matters*—the element of competition.

Competition, it will be remembered, is the key to Sartori's initial division of party systems into competitive and non-competitive systems. Given his definition of competition (see above), the non-competitive label unquestionably applies to authoritarian systems like the Polish, the East German, and the Mexican, in which one party maintains an iron grip on governmental power. It is certainly of interest that in many systems of this kind the ruling party has nevertheless chosen to allow—and, indeed, often to promote—the existence of second-class parties. But (unless one were to take the hazy 'sham'/non-sham distinction seriously) what is more informative, in real political terms, is the outcome of the second choice that such a ruling party faces. It has the choice, to put it one way, either of binding the minor parties to it as unquestioning collaborators or of allowing them publicly to adopt a competitive stance, typically by nominating opposing candidates in elections which pose no danger to the ruling party's position. The latter option does not disturb the basically non-competitive nature of the system which turns, in Sartori's terms, on the denial of 'equal rights' to opposition candidates. On the other hand, if exercised, it is an option that signally modifies the one policy fundamental to all authoritarian regimes—the policy of denying the legitimacy of public opposition to the regime. For official authorization of even hollow competition amounts to a deviation from the letter of that policy.

The element of competition, then, provides both a practicable and a significant basis for the reformulation of Sartori's definitions of one-party and hegemonic-party systems. The new definitions are as follows. In *one-party* systems political power is monopolized by a single party which may (or may not) permit the existence of minor parties, but does not allow them to act in a publicly competitive role. In *hegemonic-party* systems political power is monopolized by a single party which both permits the existence of minor parties and allows them to compete with it in elections, but never on terms that might involve the defeat of, or a lesser cost unacceptable to, the ruling party.

Applications and Qualifications

In the case of all but four of the 27 countries whose trade union movements were paid special attention in preceding chapters, Sartori's application of his categories was for the most part reasonably straightforward, and holds up well more than a decade later.

On the side of competitive party systems, he located India and Japan, decisively, and Sweden and Norway, more tentatively (and primarily on the basis of their 'longitudinal record'),[23] in the *predominant-party* category. He described the United Kingdom, the United States, New Zealand, Canada, Australia, and Austria as the only existing *two-party* systems (the last three requiring particular explanation before being settled in the category).[24] *Limited pluralism* was associated with West Germany, Denmark, Belgium, the Netherlands, Venezuela, and (subject to qualifications concerning its singularity) Israel.[25] The category of *extreme pluralism* included France, Italy, and Finland (the last also described as 'the most ... successful instance of controlled polarization'; and elsewhere, along with Israel, as 'extreme-moderate (mixed)').[26]

On the side of non-competitive systems, Sartori included (of the 27) the Soviet Union, China, and Yugoslavia among his *one-party* cases; and nominated Mexico as one of his *hegemonic-party* cases.[27]

The four (of the 27) countries omitted from this listing are Poland, Ireland, Tanzania, and South Africa. In each case. Sartori's

[23] Sartori, *Parties and Party Systems*, pp. 150, 177, 197.
[24] Ibid. 188–9.
[25] Ibid. 154–5.
[26] Ibid. 163, 310, 312.
[27] Ibid. 221, 232.

identification raises a problem. The difficulty about Poland has been mentioned above in connection with the reformulation of his two non-competitive categories. In the light of that discussion, Poland is to be located under the *one-party* heading.

The position of Ireland is confused by the fact that Sartori inadvertently identified it with two categories, predominant-party and limited pluralism. Although in each case quite emphatic, the terms of the identification vary. The allocation to limited pluralism is cursory, and without any elaboration.[28] The allocation to predominant-party, on the other hand, is at least briefly argued, and in a way that links Ireland with Sweden and Norway (as borderline cases); moreover, in a major concluding table encompassing 71 countries, Ireland is again identified as a predominant-party system.[29] The course of Irish politics since Sartori wrote (as is true also of Sweden and Norway) confirms the *predominant-party* identification.

The problem relating to Tanzania is of a different order. It was one of the newly-independent African countries which Sartori thought of as 'formless', in the sense that their institutions were in 'a highly volatile and initial stage of growth'.[30] In these circumstances, he considered it premature to apply to them the typology he had developed with an eye to more settled political systems. He therefore devised a separate typology; and identified Tanzania as bestriding two of its categories—'de facto single-party, dominant authoritarian'.[31] This amounts to a refined version of his *one-party* category, a classification of Tanzania which is readily justifiable a decade further on.

South Africa, for its part, is classified in one of Sartori's tables, without preliminary discussion, as a competitive political system of the predominant-party type.[32] Alongside this identification there is a notation (in a 'Remarks' column), 'Limited electorate': there is no similar notation against any of the other 70 states in the same table. Earlier in the book, South African electoral returns are dissected in another table which makes no reference to the nature of the electorate; and there is a passing reference elsewhere to 'white-controlled South Africa'.[33]

The point to be made is that South Africa's party system, so far as it concerns government and parliamentary elections, operates in a

[28] Ibid. 173, 182. [29] Ibid. 197–8, 200, 310. [30] Ibid. 244.
[31] Ibid. 264. [32] Ibid. 311. [33] Ibid. 193, 249.

framework that sets it a world apart from other predominant-party systems identified by Sartori, such as Sweden, Norway, India, and Japan. True, it may be described as democratic (and, indeed, as a predominant-party system within the limits of its operation), but it resembles neither the competitive liberal democracy of the West nor the non-competitive people's democracy of the East. For the democracy of South Africa is not the democracy of modern times at all, but most nearly resembles that of ancient Athens where citizenship was determined by birth, and the vast majority of adults (women, slaves, and foreigners) were excluded from the electorate. A party system formally and effectively restricted to a racial minority has no place in Sartori's competitive and non-competitive groupings; and he expressly excluded South Africa from the special typology he devised, as we have seen, for the 'formless' states elsewhere in Africa. In so far as the South African regime has repressed black political parties, it enters the sphere of the Sartori's anti-party states (see above), which he associated primarily with military regimes. But anti-party states are one of two forms taken by no-party states, political systems in which party organization is not a factor in the exercise of state power. And yet South Africa clearly does have a party system (and one which operates as a predominant-party system) so far as the ruling white minority is concerned—even though it does not touch the black majority.

South Africa thus appears to be a cross between an anti-party state (which is how the political system presents itself to the black majority) and a predominant-party system, from the perspective of the white minority. In addition, however, it has a clear affinity with a type of political system outside the range of those classified by Sartori—and that is the colony. Sartori did not seek to identify this political form—and there was no reason why he should, since (like others concerned with the identification of party systems) his eyes were on sovereign, technically independent states, whereas colonies are extensions of a metropolitan power. Nevertheless, it may be noted that South Africa has in common with the classic colony of modern times a ruling minority which is unambiguously distinguished from an overwhelmingly more numerous ruled majority by race and, most dramatically, by colour as well. More specifically, the South African situation resembles what Clark Kerr and his co-authors once categorized as 'settler colonialism' (as distinct from 'segmental colonialism' and 'total colonialism') and, indeed, applied to South Africa.'[34] Since this colonial aspect is the single most distinctive characteristic of the South African political system, it may be appropriate to label this strange anti-party/predominant-party combination a *colonial-hybrid*.

PARTY SYSTEMS AND TRADE UNION MOVEMENTS

The associations between types of party system and types of trade union movement are set out in Table 2 with reference to the 27 trade union movements given particular consideration in earlier chapters. The main points to be made on the basis of the table concern the three, more commonly encountered types of movement: in order of frequency, state-ancillary, party-ancillary, and autonomous.

1. The distinction between state-ancillary movements, on the one hand, and both party-ancillary and autonomous movements, on the other hand, involves a parallel distinction between non-competitive and competitive party systems.
2. Party-ancillary movements are associated with every type of competitive party system other than the two-party system.
3. Autonomous movements are associated with every type of competitive party system.
4. Of the competitive party systems, only one (two-party) is associated exclusively with one type of trade union movement, and that is the autonomous type.

Associations, to be sure, are one thing: causation is another. However, there does seem to be a quite pronounced causal element in the relationship between party system and trade union movement. That is suggested by the tendency for a radical change in political regime to be followed by an equally radical change in the character of the trade union movement. Take, for example, the cases of West Germany and Austria. Beginning as party-ancillary trade union movements, under weakening autocracies with parliamentary pretensions, they were later brutally transmuted into what amounted to state-ancillary movements under Hitler—before emerging, under post-war competitive political systems, as autonomous movements. The precise outcome was not ineluctable, but it would seem that the broad pattern of change (away from the

[34] Kerr *et al.*, *Industrialism and Industrial Man*, pp. 41–3.

TABLE 2. Party Systems and Trade Union Movements

System	Movement				
	Autonomous	Party-ancillary	State-ancillary	Party-surrogate	State-surrogate
Competitive					
Predominant-party	Sweden Norway Ireland	India		Japan	
Two-party	United Kingdom United States New Zealand Canada Australia Austria				
Limited (moderate) pluralism	West Germany Denmark	Belgium Netherlands Venezuela			Israel
Extreme (polarized) pluralism	Finland	France Italy			
Non-Competitive					
One-party			Soviet Union China Yugoslavia Tanzania	Poland (Solidarity)	
Hegemonic-party			Mexico		
Mixed					
Colonial-hybrid		South Africa			

state-ancillary model) almost certainly was. Thus, in Italy, Spain, and Portugal, there is the comparable but not identical spectacle of original party-ancillary trade union movements, after a dictator-induced state-ancillary interregnum, reverting to the party-ancillary form under succeeding competitive party systems. A similar pattern of change, but in the reverse direction, accompanied the post-war imposition of unequivocal one-party regimes in East Germany and other states in Eastern Europe. There, trade union movements which, pre-war, had been inclined to a party-ancillary character, were thrust firmly into the state-ancillary mould (apart from some notable, but invariably short-lived, deviations: see Chapters 10 and 11).

Inevitably, in a matter of such complexity, there are exceptions—cases in which a radical change of regime has not been accompanied by a radical alteration in the character of the trade union movement. But they are rare. One version of this kind of 'hang-over' effect is seen in the post-war history of Greece and Brazil, where the trade union movement retained a decided state-ancillary cast following more than one restoration of competitive party politics (though the pattern may have been broken in Greece since 1982 and in Brazil since 1985) as an outcome of continuing governmental use of highly intrusive legislative powers inherited, in their essentials, from dictatorships.[35] The hang-over effect has also occurred in the reverse direction, when military regimes have replaced competitive party systems—in at least the cases of Pakistan and Bangladesh, with their persisting party-ancillary trade union movements.[36] And there is, too, the unique Israeli version of the hang-over effect (see Chapter 12 and below).

Nevertheless, the exceptions are very few and limited; and, at least at the level of the distinction between competitive and non-competitive party systems, there is every sign of a strong, if not invariable, causal connection between party system and trade union movement. In particular, there is the overwhelmingly specific association of non-competitive party systems with state-ancillary trade union movements. This is a pattern which is repeated in the case of no-party political systems. Thus the spread of the state-ancillary form, accompanying the proliferation of one-party and

[35] See Kohler, *Political Forces in Spain, Greece and Portugal*, pp. 136–42; Tzannatos, *Socialism in Greece*, pp. 137–40; Juris *et al.*, *Industrial Relations in a Decade of Economic Change*, pp. 82–5; Andrade, *The Labor Climate in Brazil*, p. 93.

[36] See 'Asia's Unions', pp. 63–5.

hegemonic-party systems, has been by far the most prominent trend in the history of trade unionism since the Second World War. It seems reasonable to conclude, in the case of authoritarian systems (non-competitive and no-party), that the nature of the political system is of paramount importance as a determinant of an associated trade union movement's character.

It follows that the likely causal role of the political system is almost as clear-cut, up to a point, in the case of competitive party systems. For the nearly exclusive connection between authoritarian regimes and state-ancillary movements implies that (with exceptions in the case of the surrogates) every other type of trade union movement depends, above all, upon competitive party systems for its existence.

Beyond that level of generality, however, the issue becomes complicated. For, as the table indicates, both autonomous and party-ancillary trade union movements (leaving aside the rare surrogate movements) are associated either with all four or with three kinds of competitive party systems. At the same time, there is evidently some bunching of these associations. Thus, party-ancillary movements, strictly incompatible with two-party systems, it appears, and only occasionally linked with predominant-party systems, seem to be overwhelmingly a feature of either limited or extreme pluralism. As for autonomous movements (and in this case the table represents almost the whole population), their centre of balance tilts decidedly toward the two-party and predominant-party systems.

The overlaps, of course, are still there; and, in the table, they show up most strongly under party systems of the limited pluralism type. They point, of course, to causal factors other than the party system. So, too, does the existence of significant differences between trade union movements of the one type, a feature especially of the autonomous and the party-ancillary categories (see Chapters 9 and 13).

OTHER CONSIDERATIONS

Autonomous Movements

Six of the main autonomous trade union movements differ strikingly from both party-ancillary movements and the other main autonomous movements in the nature of their origins. The six are all in English-speaking countries: the United Kingdom, Ireland, the United States, Canada, Australia, New Zealand. Historically, five of them were heavily influenced, through migration or conquest, by the sixth (the British); and they all share the experience of being firmly established, at least among skilled manual workers, before either the emergence of a substantial Labour or Socialist party, or (as in America) the forging of fairly consistent links with a major party. Indeed, the American case apart, it was the trade unions that were largely responsible for the creation, and the early survival, of the party with which many of them in each country are now linked by way of formal affiliation. There is an obvious temptation to ascribe to these common beginnings the failure of the party-ancillary style to take root in the six countries. But an explanation wholly in these terms is highly questionable. For one thing, although the circumstances are not precisely parallel, the origins of the French General Confederation of Labour and the All-India Trade Union Congress (see Chapter 9) warn against a too ready reliance on the factor of origins. More to the point, however, a consideration of the disparate historical experience of the other, Continental, autonomous trade union movements highlights a quite different factor.

The West German and Austrian trade union movements, as we have seen, originated as party-ancillary movements. The present structures were the product of a changed post-war ideological climate, the preferences of British and American trade union leaders brought in as advisers by occupation authorities, and—one suspects—the personal ambitions of German and Austrian union leaders who had survived the Third Reich. They may, as a result, be thought of as 'artificial', blueprint creations. Nevertheless, they have lasted as plainly autonomous movements. The contrast to be drawn is with Italy. There are the same factors of a party-ancillary tradition, dictatorship, military defeat, and Allied occupation. In Italy, too, there was the initial resolve to create a single major trade union confederation in the autonomous mould (see Chapter 9). But, in that case, the blueprint did not work; and the Italian movement reverted to its tradition. There is one critical factor which distinguishes these cases. It is the post-war presence in Italy, but not in West Germany or Austria, of a powerful Communist party which ensured the polarization of both trade union and party politics once the pressures born of war and occupation had eased.

The three Scandinavian countries raise another kind of problem. As was the rule elsewhere in Continental Europe, socialist parties were a dominant force in the early years of trade unionism in Sweden, Norway, and Denmark. Nevertheless, the trade unions of these three countries have continued throughout to display the characteristics of autonomous movements—unlike, for example, the neighbouring party-ancillary movements of Belgium and the Netherlands. The relative strengths of different Communist parties do not account for this difference. Belgium, the Netherlands, and the Scandinavian countries have all failed to provide congenial soil for such parties. What does, however, distinguish the Scandinavian countries is the absence of a substantial Roman Catholic community. In Belgium and the Netherlands, as elsewhere, Catholicism has historically provided the main source of a polarizing force on the right wing of trade union politics—the Christian confederations and their associated unions.[37]

The fourth Nordic nation, however, does have a strong Communist party. Indeed, Finland boasts the 'third major Communist party of Western Europe'.[38] But this party, in contrast to its Italian and French counterparts, ceased playing a genuinely polarizing role in both the trade union movement and the political system at large in the mid-1960s. Ever since then it has been a frequent partner in coalition governments. At that time, too, it converted the Finnish trade union movement from a party-ancillary to an autonomous movement by opting to abandon its own trade union confederation and merge into a confederation within which its members would be, and have remained, the major opposition to the dominant Social Democrats.

The six Continental autonomous movements share the party-oriented nature of their origins with the Continental party-ancillary movements (France apart)—but not, of course, their category. In contrast, they share their category with the English-speaking autonomous movements—but not the nature of their origins. It follows that there can be no decisive connection between historic origins and contemporary category. The crucial connection involves, instead, the distinctive feature that emerged in the discussion of the Continental cases. For what the six English-speaking autonomous movements share with their Continental counterparts is the fact

[37] See Fogarty, *Christian Democracy in Western Europe*, esp. ch. 15, 16.
[38] Tannahill, *The Communist Parties of Western Europe*, p. 8.

that they inhabit a competitive political system in which there is no party both interested in and capable of polarizing trade union organization. The outcome for trade unionism in political systems where such parties do exist is considered in the next section.

Party-ancillary Movements

The inspiration for the formation of confederations in trade union movements of the party-ancillary type has come, usually, from specific political parties (see Chapter 9). Major exceptions to the rule are the French General Confederation of Labour and the All-India Trade Union Congress—though, eventually, each came to be controlled by a single party (see Chapter 9). Some such association with a single party, on a continuing basis, is typical of confederations in party-ancillary movements. The nature of the connection, as we have seen, varies from tight party control to a loose alliance of equals. Beyond that, there is the relatively rare exception of the French Democratic Confederation of Labour with its fairly determined syndicalist posture—though even it, as we have seen, has flirted for a time with a party. A slightly less uncommon exception involves a continuing association with *two* parties, a situation which is likely to be reflected in agreed power-sharing arrangements between the parties' adherents within a single confederation: the Confederation of Venezuelan Workers and, if not quite so unambiguously, the Italian General Confederation of Labour provide examples of such arrangements. In comparison with autonomous trade union movements, however, this kind of intra-confederal coexistence is usually limited and partial in the case of party-ancillary movements.

Party-ancillary movements are a function of centrifugal pressures involving an intensity of ideological antagonism sufficient to override centripetal, unifying pressures arising from common interests defined by such things as occupation, industry, or class—or, as in India during the 1930s, by a struggle for national independence. This kind of polarization within a trade union movement reflects, and appears to be primarily a function of, polarization within the political system at large. And in this respect (the special case of South Africa apart), two features are common to the party systems associated with the party-ancillary movements examined in Chapter 9. The first is an electorally substantial, mass-membership party of the Social Democratic or Labour kind, with a long-standing, clear and active interest in trade union organization (a characterization, it should be noted, which is applicable to the Indian National Congress). The second is an electorally substantial party, or parties, either to the left or the right, or both, possessing a similarly direct interest in trade union organization. These are the polarizing parties, which are typically, if to the left, Communist parties and, if to the right (in the European cases), Christian Democratic parties.

But there are different degrees of polarization. Inter-confederal relations within party-ancillary movements range, as we have seen, from fiercely competitive (high polarization) to strongly co-operative (low polarization). In this respect the cases considered in Chapter 9 (again, South Africa apart) fall into three distinct categories which seem to match well enough the corresponding aspect of their respective party systems.

Belgium, the Netherlands, and Venezuela, with party systems of the limited pluralism type, also have trade union movements marked by notably low levels of confederal competition. What distinguishes these three cases from the others, with higher levels of competition, is the presence on the left of a Communist party which is weak, in both the country and the unions; and the presence on the right of a comparatively strong party (or parties) with a religious orientation. The result, in Belgium, is the Confederation of Christian Unions, the largest of three confederal bodies. In the Netherlands, the merger of Socialist and Catholic confederations in 1976 has left the conservative and formerly Protestant confederation, now strengthened by the accession of disaffected Catholic unions, to sustain the competitive element on its own. In Venezuela, the Social Christian party both supports a confederation of its own and acts as junior partner in the two-party coalition which controls the main confederation.

France and Italy share more polarized (extreme pluralism) party systems and higher levels of confederal competition within their trade union movements. They also have in common a Communist party of major electoral significance. In each case, the Communist party is associated with the largest of the trade union confederations. In each case, too, there is a Catholic-oriented confederation—although the Italian Confederation of Workers' Unions is far more substantial than the French Confederation of Christian

Workers (*Maintenu*); and, also unlike its French counterpart, is linked to a party, the Christian Democratic party, of consistent relevance to governmental coalitions. On the other hand, the degree of polarization in each case is not the same. The Italian confederations, as we have seen, have displayed a strong inclination towards co-operation, while inter-confederal competition is demonstrably more intense in France. Almost certainly, this owes something to differences relating to the respective Communist parties. The French party has a more aggressive and exclusivist tradition than the Italian party, which has a quite contrasting history of co-operation, at various levels and times, with other parties on the left.[39] In addition, the French party, unlike the Italian, has retained tight control over its trade union confederation, and this has helped insulate the latter from the industrial (as distinct from political) pressures which have played a major part in pushing its Italian counterpart towards more co-operative policies.

Differences apart, however, a notable factor common to the French and Italian cases would seem to be the existence, not merely of an electorally strong Communist party, but of a Communist party which is generally excluded (whether by itself or by others) from governing coalitions. The likely significance of this factor is suggested, in one way, by the experience of Finland where, as we have seen, the absorption of a major Communist party into the politics of coalition-making has been accompanied by the confederal unification of the trade union movement. It is suggested in another way by the experience of Italy itself. During 1944–7 the Communists, Socialists, and Christian Democrats there were joined both in governmental coalition and in the one trade union confederation: the Christian Democrats' decision to exclude the others from the coalition was followed by the organizational splintering of the union movement.

India, with a predominant-party system, has a trade union movement which is remarkably fragmented in structural terms, and in which confederal competition appears to be highly intense. Some of this intensity may well be a function of the violent edge to politics in a society with intractable communal and caste cleavages. Otherwise, however, there is the fact that trade union organization in India is caught up in the swirl of parties around the skirts of the predominant Congress party. The other parties with a direct

[39] See Tannahill, *The Communist Parties of Western Europe*, pp. 221–4.

interest in the organization of urban employees range, as in Italy, across the political spectrum. To Congress's right, with its own trade union confederation and expressing official sentiments about labour and trade unions which are reminiscent of European Christian Democratic parties,[40] there is the Jana Sangh, the major right-wing party. To Congress's left, there is the Socialist party, the original Indian Communist party, and a number of breakaway parties. This splintering on the left, as we have seen, is directly reflected in the structure of trade union organization—thus adding a sectarian dimension to confederal competition which may be expected to sharpen rivalries and heighten competition. In any case, tight party control, which seems to be generally characteristic of Indian confederations, helps ensure the dominance of political considerations in their relations with each other.

Among party-ancillary trade union movements, the South African case is unique. Like the peculiar political system this movement inhabits, its explanation begins and ends with the stark pervasiveness and overwhelming potency of the racial issue in South African affairs.

State-ancillary Movements

Trade union movements of the state-ancillary variety, in comparison with the less prevalent types, are explained quite simply. They owe their genesis *invariably* to an authoritarian government. What this usually means is that they are a product of non-competitive party systems; and otherwise of no-party systems. There is virtually no ambiguity about the connection. The structure and management of state-ancillary movements are in their essentials government determined. For the greater part, whether one-party or not, the governments have opted (to fasten on the most readily observed indicator) for monolithic, all-inclusive trade union confederations. Occasionally, for a time, as in Brazil, Chile, and Ghana, regimes have opted for no central confederation at all.[41] Mexico, as we have seen, provides a more durable and more notable exception to the rule with a pluralistic confederal structure matching the ostensibly

[40] See Bhandari, 'Bharatiya Jana Sangh', pp. 135–6; Sirsika and Fernandes, *Indian Political Parties*, pp. 168–73.

[41] See Cordova, *Industrial Relations in Latin America*, p. 32; Schlagheck, *The Political, Economic, and Labor Climate in Brazil*, pp. 68 ff.; Damachi *et al.*, *Industrial Relations in Africa*, pp. 145–6.

pluralistic character of the Mexican party system itself. But hegemonic-party systems in general, as exemplified by Egypt, Indonesia, and Singapore, tend to prefer the monolithic solution.[42] On the other hand, there is Turkey as a contemporary, if much less adventurous, emulation of the Mexican case;[43] and confederal organization in Brazil appears to have moved in the same direction since civilian government was last restored there.

The Surrogate Movements

Of the three surrogate cases examined in Chapters 11 and 12, the Polish case alone has known predecessors, all of them on the African continent. The essential similarities with the classic colonial model are unmistakeable in the circumstances which produced the Independent Self-Governing Trade Union Solidarity as a legal opposition to the official (state-ancillary) trade union apparatus during 1980–1. On the one hand, there was the entrenched authoritarian government unequivocally identified with a dominating foreign power. On the other, there was the nationalist movement, based on a sense of ancient nationhood, which identified Russia as the traditional enemy and the Catholic Church as the spiritual manifestation of the Polish nation.

The circumstances of the Japanese case are somewhat more sedate, as we have seen. A trade union movement, fragmented along political lines, inhabits a predominant-party political system in which two left-wing parties with significant electoral support are condemned to seemingly permanent opposition. Moreover, the two parties, both non-Communist (the absence of the Leninist tradition is important), have no mass membership pretensions, unlike their European and Indian counterparts. Instead, like most Japanese parties (owing partly to their origins and partly to an unusual electoral system) they are highly 'fractured' and decentralized in electoral—though not in parliamentary—terms; and they rely for their organizational and financial underpinning less on members than on supporting interests, especially at the local level for electoral purposes.[44] Lack of competition for the role of patrons of

[42] See Ayubi, *Bureaucracy and Politics in Contemporary Egypt*, pp. 451–3; Reeve, *Golkar of Indonesia*, pp. 331–2, 349–51.

[43] See Blum, *International Handbook of Industrial Relations*, pp. 560–90.

[44] See Sartori, *Parties and Party Systems*, pp. 90–4.

two parties without prospect of governmental power reinforces the position of trade union leaders in their relations with the Socialist party and the Democratic Socialist party.

The Israeli case, the one state-surrogate instance, has something of importance in common with each of the party-surrogate cases. On the one hand, the Histadrut shares with the Japanese confederations the present context of a competitive party system (which Sartori, while identifying it as limited pluralism, describes as 'a most baffling case ... a microcosm of all the conceivable complexities').[45] It shares with Poland's Solidarity, on the other hand, the essential nature of the political system in which it originated—an authoritarian regime (if perhaps less comprehensively so than in the Polish case) identified with an alien power. The Histadrut's state-surrogate characteristics, as we have seen, were most fully developed under that regime. But then there are the singular features. A people following a dream, seeking a refuge, the Jews who were to create the state of Israel came mainly from European and urban settings. They came to an impoverished Middle Eastern country that was governed by a metropolitan power and occupied by an ethnically and religiously distinct population. But, of all people, the Jews had mastered the art of surviving as a minority in a hostile environment. Behind this were hard lessons learned through centuries of segregation and persecution, self-reliance and inventiveness. As a result, while capable of splintering into a myriad political parties, they also had the resolve, the talent, and the experience as a community to organize and develop for themselves the sophisticated services and support systems of a modern society. The Histadrut was both the vehicle and the demonstration of this.

[45] Sartori, *Parties and Party Systems*, p. 151.

15

THE THEORY

The theory of national trade union movements that emerges from the preceding chapter inevitably focuses on just three of the taxonomic categories developed earlier. State-ancillary, party-ancillary, and autonomous are the categories that matter: they jointly enclose virtually all contemporary trade union movements. (The two surrogate categories, moreover, depict types which are not only extremely uncommon but, historically, have mostly tended to be short-lived.)

The theory deals with two levels of causality. The first has to do with the broad nature of the political system. The distinction to be made at this level is between states with competitive party systems, on the one hand, and states with either non-competitive party systems or no-party systems, on the other hand.[1] This distinction identifies the primary determinant of a trade union movement's character.

The key propositions are as follows. (1) Non-competitive party and no-party systems almost invariably produce trade union movements that fall into the state-ancillary category. (2) Competitive party systems almost invariably produce trade union movements of *either* the autonomous type *or* the party-ancillary type. The marginal qualification ('almost invariably') of the causal connection in each of these propositions, as in later ones, acknowledges the rare exceptions to the rule—surrogate products of both competitive and non-competitive systems; and trade union movements subject to a hang-over effect following a systemic change from competitive party system to non-competitive (or no-party), or vice versa (see Chapter 14).

The nature of the political system, as primary cause, possesses the status of sole and sufficient cause only in the case of state-ancillary trade union movements. State-ancillary movements are *invariably* a

[1] This formulation, it will be observed, ignores the mixed (competitive/non-competitive) 'colonial-hybrid' system found in South Africa (see Chapter 14). That case, however, must be treated as *sui generis*.

product of either non-competitive party systems or no-party systems.[2]

One implication of this conclusion about state-ancillary movements—given the enormous variations evident in the economic and social circumstances of countries blessed with non-competitive party systems or no-party systems—is the causal primacy of the political, as against the economic and the social. Moreover, this implication is sustained in relation to autonomous and party-ancillary movements, although their aetiology is less straightforward than that of their state-ancillary counterparts.[3]

The basic requirement for both autonomous and party-ancillary trade union movements is a competitive party system. But they depend upon a secondary feature of that system for their precise differentiation as either autonomous or party-ancillary movements. This secondary feature is the presence (or absence) of one or more polarizing parties.

A polarizing party, for present purposes, has the following attributes.[4] First, it is electorally significant, with substantial parliamentary representation. Second, it is identified with either the extreme left (typically Communist) or the strongly conservative (not Fascist) right. Finally, if a party of the right, it actively pursues an explicit and discrete organizational interest in trade unionism; and, if a party of the left, it is both in a permanent parliamentary minority and persistently unable, or unwilling, to participate in coalition governments.

The key propositions are as follows. (1) Competitive party systems that *include* a polarizing party (or parties) almost invariably generate party-ancillary trade union movements. (2) Competitive

[2] This, of course, includes hang-overs under succeeding competitive party systems (see Chapter 14).

[3] It may be recalled, in any case, that the limitations of a primarily socio-economic explanation of national trade union movements have already been suggested by the inadequacies (especially in relation to the category of developing countries) of the three-world formula as a classificatory scheme: see Chapter 8.

[4] The conception of a polarizing party, set out here, is not to be confused with Sartori's 'anti-system' party (see Sartori, *Parties and Party Systems*, pp. 132–4), which also has a polarizing role. The difference between the two notions is, in the first place, a matter of context: anti-system parties polarize whole political systems, whereas polarizing parties need do no more than polarize trade union movements. In the second place, the difference is a matter of the specific type of party involved. For example, while Communist parties figure in both categories, Fascist and Christian Democratic parties do not: Fascist parties figure only as anti-system parties, Christian Democratic parties only as polarizing parties.

party systems that *lack* a polarizing party almost invariably generate autonomous trade union movements.

On both sides of this dichotomy the polarizing party is the critical factor, but for different reasons. Its role is unambiguously positive in the case of party-ancillary movements: it creates them. In the case of autonomous movements, the absence of the polarizing party confers the creative role, by default, on other factors. The question in this case is: what other factors?

The first point to be made is that any other, creative factors are bound to be intrinsic to the trade union movements concerned. The contrast here is with state-ancillary and party-ancillary movements, both of which are shaped predominantly by extrinsic influences operating through governments or parties. It is the comparative independence of autonomous movements from extrinsic influences which almost certainly explains the greater variety evident (see Chapter 13) in their structure, functions and formal external relationships.[5] Nevertheless, along with this diversity, there is one tell-tale regularity of behaviour which further distinguishes autonomous movements from both their state- and their party-ancillary counterparts. That regularity in the behaviour of autonomous trade union movements is an emphasis on collective bargaining as the principal method of union action.

Collective bargaining and its implications are the focus of Hugh Clegg's pathbreaking book, *Trade Unionism under Collective Bargaining*.[6] The theory he advanced there is that, in certain circumstances, it is the specific structure (the compound of a range of varying 'dimensions') of collective bargaining in different countries which chiefly accounts for the shape of significant features of national trade union movements, including membership

[5] It is to be emphasized that the independence of autonomous movements from extrinsic influences is a relative matter. They are independent, that is to say, in comparison with state- and party-ancillary movements, and with particular reference to states and to parties as sources of extrinsic influence. The comparison, on the other hand, may be extended to take in a third major source of extrinsic influence—but, this time, one that tends to be more potent in relation to autonomous movements than their state- and party-ancillary counterparts. That source is private employers. The key variables involved are the structure of organization among employers and the structure of management, together with employers' attitudes: see Clegg, *Trade Unionism under Collective Bargaining*, pp. 106–8. These factors, as Clegg demonstrated, have critically influenced the dimensions of the collective bargaining systems in which autonomous movements operate.

[6] For a concise and definitive analysis of the concept of collective bargaining, see Clegg, ibid. 5–8.

density, aspects of union government, the role of workplace organization, strike patterns, and attitudes to industrial democracy.[7] But, crucially, this theory is not applicable wherever collective bargaining is used. It applies only in countries where collective bargaining is unequivocally the *principal* method of trade union action. The point is underlined in the six national cases Clegg chose to examine in the process of establishing his theory: Australia, Sweden, the United Kingdom, the United States, West Germany, and France. The theory, as he demonstrated, applies in all cases but the last. It does not apply to France because there, alone among the six cases, the trade union movement is 'at least ambivalent between political action and collective bargaining as methods of trade union action', with the result that collective bargaining is 'rivalled, if not surpassed, by political action' as a preferred method.[8] The French exception, it will be noted, is consistent with the autonomous/party-ancillary distinction of the present work: Clegg's theory, that is to say, discriminates neatly between the five autonomous trade union movements and the one party-ancillary movement in his sample. Moreover, there is every reason to believe that his theory discriminates with equal precision between autonomous and party-ancillary movements in general.

Thus, it is apparent that a decisive emphasis on the method of collective bargaining is one thing that emerges in the absence of a polarizing party. But there is more to it than this. For, plainly, an emphasis on a particular method of action implies an emphasis on goals that are at least potentially achievable through that method. The point about this, in turn, is that the objectives attainable by way of collective bargaining are strictly limited.

Collective bargaining, hinging as it does on direct dealings between unions and employers, is necessarily focused on a specific field of concern, the parameters of which are defined by the range of employers' decision-making authority. The goals of collective bargaining are restricted correspondingly. 'Its subject-matter', as Clegg simply stated it, 'is terms of employment.'[9] The bench-mark is the objectives of political action, the other main trade union method of action. And the goals potentially attainable by way of political action are infinitely more expansive—ranging as they may all the way from terms of employment, through social reform, to the revolutionary transformation of society.

[7] Clegg, *Trade Unionism under Collective Bargaining*, esp. pp. 8–11, 118–9.
[8] Ibid. 116, 119.
[9] Ibid. 5.

It follows that by emphasizing collective bargaining over political action, autonomous trade union movements are also signalling an overriding concern with a narrow range of issues—specifically, the pay and working conditions of their members. It was this kind of limited concern that, on the one hand, Lenin belittled as 'trade-union consciousness'[10] and, on the other, Selig Perlman celebrated as the 'philosophy of organic labor'.[11] Both of them, nevertheless, were at one in their perception of this narrow approach as an accurate reflection of the 'spontaneous' objectives (Lenin's term),[12] the ' "home-grown" ideology' (Perlman's phrase),[13] of workers themselves. They agreed, too, that the only way trade unions were to be diverted from this approach was by the intervention of outsiders—whether they were Lenin's benign 'professional revolutionaries',[14] Perlman's malign 'intellectuals'[15] or (though neither of them lengthened the list in this way) the others, the capitalists, *apparatchiks,* and colonels, who have shaped the official concerns of specific trade union movements by acting through governments or parties. Their joint conclusion, to put it another way, was that trade unions *left to themselves* will stress, above all else, goals which match the limited horizons of the bulk of their members.

The empirical accuracy of this conclusion is no longer a debatable issue, as attitude surveys and worker revolts, both West and East, have repeatedly testified. In other words, what emerges in the absence of both an authoritarian state and a polarizing party is a trade unionism which (with all the distortions that arise in leader–member relationships) tends, on balance, to be more responsive to the pressures emanating from its members than to those emanating from outside influences. Perlman, as we have seen, used the highly suggestive term 'organic labor' in this connection. He has a point. To the extent that their purposes appear to be, by and large, expressive of their members' desires, autonomous trade union movements are arguably the 'natural' form of trade unionism.

[10] Lenin, 'What is to be Done?', p. 375.
[11] Perlman, *A Theory of the Labor Movement*, p. 254.
[12] 'What is to be Done?', p. 384.
[13] *A Theory of the Labor Movement*, p. 6.
[14] 'What is to be Done?', p. 464.
[15] *A Theory of the Labor Movement*, p. 9.

APPENDIX

The Kerr *et al.*, Millen, Davies, and Cella and Treu Typologies (see Chapter 8)

Kerr et al. *(1960)*

The central thesis of Kerr, Dunlop, Harbison, and Meyers was that the world was moving towards complete industrialization under the leadership of 'industrializing élites', of which there were at the time five 'generalized types': respectively, the middle class, dynastic leaders, colonial administrators, revolutionary intellectuals, and, finally, nationalist leaders.[1] Each of these élites, it was argued, gave rise to a specific kind of 'labor organization'. In other words, trade union movements were basically divisible into five types according to the industrializing élite with which they were associated. The characteristics of each of the five types were elaborated in terms of seven features, including ideology, functions, involvement with workplace rules, inter-union competition, structure, sources of funds, and sources of leadership.[2] Table 3, reproduced from *Industrialism and Industrial Man*,[3] summarizes the nature and connections of these categories.

Millen (1963)

Bruce H. Millen, in his approach to the taxonomic issue, rejected the two assumptions common at the time he wrote, especially among his American compatriots. He denied that 'economic' and 'political' unionism were mutually exclusive in practice, and he denied that American unions were a pure case of the former. He defined economic unionism basically with reference to the use of collective bargaining. His definition of political unionism was more complicated. It involved six 'characteristics'.[4] Four of these were: the continual involvement of union leaders in 'direct political

[1] Kerr *et al.*, *Industrialism and Industrial Man*, pp. 32–51. An earlier, and more limited, version of this thesis was advanced in Dunlop, *Industrial Relations Systems*, pp. 317–34.

[2] Kerr *et al.*, *Industrialism and Industrial Man*, pp. 193–201.

[3] This table is from the second edition (1964), and is more elaborate than the corresponding table in the first edition (1960).

[4] Millen, *The Political Role of Labor in Developing Countries*, pp. 9–10.

TABLE 3. Worker Organizations and the Élites

Industrializing Élite	Middle Class	Dynastic	Revolutionary-Intellectuals	Colonial Administrators	Nationalist Leaders
Ideology	Reformist	Class-conscious and revolutionary except for a minority.	Preserve the true revolution.	Independence	Nationalism
Functions of workers' organizations	Regulates management at the local and industry level. Independent political activity accepted. Does not challenge the élite.	Social functions at plant level; little constraint upon management. Provides minimum industry conditions by legislation. Political activity challenges the élite.	Instrument of party to educate, lead workers and to stimulate production. No political activity except through the party.	Largely a part of the independence and nationalist movement.	Confronts the conflicting objectives of economic development and protection of workers.
Role of labor organization at work place	Active role under established procedures. Regulate management.	Little role, competitive with works councils. Co-operate with management.	Little direct role; influence through party. Increase productivity.	Little role, force for nationalism and independence.	Little direct role; influence through government tribunals; increase productivity.
Division of authority on rule making	Pluralistic with workers, management and state having an active role.	State and management dominant.	Party and state, with management and labour organizations as instruments.	Manager dominant with support of mother country.	Nationalist state and enterprise managers.
Broad or detailed systems of rules	Detailed regulation at the work place primarily through collective agreement.	General rules at the industry level with management free at the work place.	Detailed regulation at industry and work levels prescribed by the state.	General rules prescribed by the state with management free at the work place.	General rules prescribed by state with management often free at the work place.

Competition among workers' organizations	Exclusive representation and keen competition. Some rivalry between plant and industry levels over allocation of functions.	Limited rivalry at the plant level and the distribution of functions between the local and industry levels. No exclusive representation.	No rivalry or competition allowed.	Divided by ideological, tactical, regional and personal leadership factions.	Tendency for consolidation among organizations recognized as loyal by nationalistic élite. Advantage over those not so recognized.
Structure of worker organizations	A variety of structural forms. Confederations not centralized. Organizations perform a wide range of functions.	Relatively large number of industrial unions. Centralized confederation often limited by rival confederations. Unions perform narrow range of functions.	A few industrial unions. Centralized confederation. Organizations perform a narrow range of functions.	A wide variety of structures. Organizations not well developed, often personal.	Tendency towards industrial unions with one confederation acceptable to élite.
Sources of funds	Substantial resources secured by regular dues; regulatory functions require administrative organizations and large budgets.	Meagre resources from irregular dues payments and indirect government allowances. Financial success not highly regarded by workers' organizations.	Substantial resources secured by assessment of all workers; financial resources present no problem with support of regime.	Meagre funds often raised outside workers' organizations.	Funds often secured indirectly from government in addition to meagre dues. Officers receive other salaries.
Sources of leadership	The ranks through lower levels of workers' organizations. They have an established career.	Intellectuals and those ideologically oriented toward political activity. The leaders' income position is often insecure.	Reliable party leaders with experience in work organizations. They have an established career.	Nationalist and independence leaders. Intellectuals with a personal following.	National leaders and intellectuals except where confined to manual workers.

Source: Kerr *et al.*, *Industrialism and Industrial Man*, 2nd edn., pp. 204–7. Reprinted by permission, Harvard University Press.

work'; 'very broad' leadership goals; 'frequent use of direct mass action' for 'nonindustrial' aims; and a requirement of 'ideological conformity' on the part of leaders. The fifth characteristic he described as 'a marked tendency towards "movementism" '. This, as elaborated, amounted to union involvement in a political party; and, in effect, overlapped with the sixth characteristic in which a union, although invariably only temporarily, 'closely resembles a political party'.

Millen then went on to 'offer' a 'spectrum' of seven categories, which ranged, in terms of the national examples he used, from the United States at one extreme to the Soviet Union at the other. He introduced each category with an illustrating case or cases followed by an outline of the factors involved, as shown in Table 4, which is derived from his text.[5]

Millen relied, in his spectrum, explicitly on three criteria: the unions' relations with the state, their relations with political parties, and their use of collective bargaining. At the same time, he clearly accorded collective bargaining a subordinate status by treating it as a variable dependent on the other two criteria.

Davies (1966)

The crucial taxonomic question for Ioan Davies was 'what kind of relationship do unions . . . have with their governments?'[6] He himself answered the question in two different ways. One figured in a chapter dealing with African one-party states where the trade unions 'tend to become agencies for increasing national production'; the contrast he drew was with European unions which, as 'agencies for greater consumption . . . act as pressure groups on governments and employers'.[7] His other answer was more complicated. It approached the issue by way of 'the developed world' and what he described as 'three situations'.[8] These are set out in Table 5, which is derived from Davies's text.[9]

These variations between the union movements of developed countries, Davies considered, 'simply illustrate that everywhere the governments either control unions or have evolved means of coming to terms with them while introducing certain restrictions

[5] Millen, *The Political Role of Labor in Developing Countries*, pp. 11–14.
[6] Davies, *African Trade Unions*, p. 135.
[7] Ibid. 152. [8] Ibid. 136. [9] Ibid.

TABLE 4. Trade Union Movements: Millen

Illustrating Cases	Trade Union Characteristics
United States	Dominating emphasis on collective bargaining; some 'political action'; unions 'free and independent of government and political parties'.
'North European countries—typified by [the unions] of Scandinavia'	Collective bargaining important; unions have 'functional links' with political parties; historical use of 'political strikes' becoming rarer; despite closeness to parties, unions 'can . . . be considered "free and independent" '.
Israel (specified as the sole case)	Collective bargaining 'commonly used' but 'not dominant'; the central union body (Histadrut) is 'as much an expression of political drive as it is of trade unionism'; close union ties with parties; Histadrut, as a 'power centre', is not 'a captive'—and 'is certainly philosophically closer' to category 1 than to category 7.
Italy, France, India, Sri Lanka ('Many of the Latin American and Caribbean countries would also seem to fit.')	Dominating emphasis on 'ideological concepts'; unions 'highly politicised' within 'a competitive political system'; strikes used for 'political purposes'; links with parties are 'the norm' with 'varying . . . party influence on trade union matters'; union independence varies within countries (depending on which parties involved) as well as between countries.
Middle East (Israel and Lebanon apart)	Unions subject to 'stern control' by government; collective bargaining 'limited by both the nature of the economy and the restraints imposed by government'.
Africa[a]	(*a*) Unions in 'many' cases subject to government control, but 'the degree of restraint is probably somewhat less than that exercised by Middle East governments'.
	(*b*) In some cases 'a more dynamic type of union is dominant'—'highly political', and associated with a mass ruling party; collective bargaining, if not usually 'emphasized . . . is not neglected'; unions, although 'now coming under increasing government control', have 'retained a voice in high councils and varying amounts of influence on over-all policy'.
Sino-Soviet bloc	'Collective bargaining is clearly subservient to the goals of the state, the trade unions are clearly subordinated to the Communist Party, and the question of union autonomy is irrelevant.'

[a] As indicated, Millen's explanation of this category effectively divided African union movements into two distinct types.

and obligations'.[10] As for *developing* countries, his initial thought was that 'the effect and extent of government intervention may be equally varied'.[11] Subsequently, he qualified this with the remark that governments in such cases were 'particularly prone to impose strict limits on the activities of unions'.[12] It is to be noted, however, that despite Davies's professed reliance on union–*state* relationships as the discriminating factor, he brought in union–*party* relationships in this role when it came to establishing his three sub-categories concerned with the trade union movements of 'industrial capitalist states'.

Cella and Treu (1982)

The classification of trade union movements devised by G. Cella and T. Treu is based on eight 'dimensions of union behaviour' which relate, in turn, to 'union structure, density, degree of centralization, workshop organization, types of union action, relationship with political parties . . . industrial conflict and relationship with the state and political system'.[13] These dimensions are each sub-divided into 'models'/'categories'/'patterns' (ranging in number from three to six in different cases) with which Cella and Treu identified a specific country or countries—the 'developing countries' and 'socialist countries' being treated as virtually undifferentiated groupings.[14] From this, Cella and Treu derived (without, it must be said, explaining how) five 'models of trade union movements' which they labelled: 'opposition, business (or domestic), competitive, participation, state-sponsored'.[15] Their table summarizing the connection between this classification and the various dimensions is reproduced (including its footnotes) as Table 6.[16]

Apart from this table, the only elaboration of these five 'models of trade union movements' which Cella and Treu provided (other than references to relevant countries) was as follows.[17]

[10] Davies, *African Trade Unions*, p. 136. [11] Ibid. [12] Ibid. 137.
[13] Cella and Treu, 'National Trade Union Movements', pp. 218–19.
[14] Ibid. 219–20. [15] Ibid. 221.
[16] Ibid. 222. This table is from the second edition (1985). The table in the first edition (1982) differed in two respects. Japan was included alongside 'USA' under the 'Business unionism' head; and the second footnote was confined to the first six words of the reproduced version. [17] Ibid. 221.

TABLE 5. Trade Union Movements: Davies

Illustrating Cases	Trade Union Characteristics
Spain, Portugal	Independent trade unions banned by the state, which has 'attempted to replace them with national union fronts' with 'little industrial importance and . . . political effectiveness (the old unions have simply gone underground)'.[a]
'[M]ost Communist countries'	'[S]ome unions are merged into the party front unions, but the whole trade union movement, however closely controlled through legislation, is able to play an influential part in the machinery of industrial negotiations and management.'[a]
'[M]ainly in the industrial capitalist states'[b]	(*a*) Trade unions having 'close working relationships' with the 'ruling parties': Sweden, Norway, Israel, and 'Britain under Labour'.
	(*b*) Trade unions, linked with an opposition party, 'have to accept severe state limitations on their activities': Germany, Australia, Japan, France.
	(*c*) Trade unions 'are not controlled by the state . . . do not have alliances with political parties, and . . . there is no voluntary control of wage claims (as in Sweden)': United States, Canada—but in these cases 'the trade unions themselves are highly bureaucratic in structure and have come to fairly close working arrangements with the state bureaucracy (as opposed to individual parties)'.

[a] This is the full extent of what Davies had to say by way of explaining the nature of these two categories.
[b] Under this heading (having specified in an introductory sentence that it involved only one 'situation') Davies referred to 'situations' in the plural.

TABLE 6. Trade Union Movements: Cella and Treu

	A	B	C	D	E	F	G	H
PATTERNS OF UNIONISM	Union structure	Union density	Coll. bargaining and union structure	Workplace organization	Union action	Industrial conflict	Relations with polit. parties	Relations with polit. system
Opposition unionism (France: CGT; Italy: CGIL); some developing countries	horizontal	low	bipolar or medium central	none (single) non-union	opposition	medium low	dependence	repression *(laissez-faire)*
Business unionism (USA)	craft industrial general enterprise	low	decentralized	single-channel union	coll. bargaining (coll. barg. + partnership)	high medium	none (or occasional)	*laissez-faire*
Competitive unionism (Britain; Italy)	all models	medium	bipolar or medium central	single-channel union	coll. barg. + occasional partnership	high medium	inter-dependence	pluralism
Participatory unionism (FR Germany; Austria; Scandinavians)	occupational industrial territorial	high	centralized (bipolar)	dual	coll. barg. + participation + partnership	low	inter-dependence	neo-corporatism

State-sponsored unionism (Socialist countries)	occupational high industrial territorial		centralized	single-channel union	partnership	none	dependence	integration

Notes

Models in brackets are not typical.
Japan is in some respects atypical: some characters [*sic*] of low conflict, participation, and partnership at the enterprise level are typical of the *participatory* pattern (even if decentralized bargaining structure, lack of corporatist relating with political parties and political systems are more typical of *business unionism*).

Source: Cella and Treu, 'National Trade Union Movements', in Blanpain (ed.), *Comparative Labour Law and Industrial Relations*, 2nd edn., p. 222. Reprinted by permission, Kluwer Law and Taxation Publishers.

Opposition unionism . . . is mainly organised on a territorial basis, since its main resource is represented by the general mobilisation and organisation of the labour force on political grounds and occasions.

Business unionism . . . is identified for its mainly economic objectives, pursued strictly through collective bargaining, outside stable political initiatives, and by relying mostly on direct organisation at the workplace.

Competitive unionism: its objectives are broader; they include basic socio-economic reforms and are pursued by initiatives both on the economic and political fronts, often highly conflictual, with close but not necessarily institutionalised relationship [*sic*] with the political system . . .

Participatory unionism operates in neo-corporatist environments and is an essential actor in the system of economic and political tripartite bargaining which characterises these environments; this implies high institutional involvement both at the enterprise and labour market level . . .

State-sponsored unionism is related to a system of industrial relations strictly controlled by the state and operates towards objectives defined at the state level with public or quasi-public functions, which exclude proper bargaining and conflictual action . . .

Cella and Treu then went on to consider the question of 'which dimensions are most decisive or comprehensive' as indicators of the various 'models'—and concluded, rather brusquely, that 'the most decisive . . . are union density, workplace organisation, relations with political parties and with the political context of industrial relations'.[18] This conclusion, in view of their elaborations (quoted immediately above) of the five 'models', is somewhat surprising. The 'dimensions' specified in the elaborations suggest a quite different conclusion. For one thing, those that can be regarded as more or less explicitly mentioned in the elaborations do not include 'union density' which Cella and Treu included among 'the most decisive' dimensions. On the other hand, three of the dimensions which they relegated to only a minor defining role ('union structure', 'types of union action' and 'industrial conflict') *are* effectively referred to—the first in one elaboration (opposition), the second in all five elaborations, and the third in two (competitive and state-sponsored).

Of the three remaining 'decisive' dimensions ('workplace organisation', 'relationship with political parties' and 'relationship with the state and political system'), the first is specified in only one elaboration (business), the second *perhaps* in one (competitive, on a

[18] Cella and Treu, 'National Trade Union Movements', pp. 221, 223.

generous interpretation of 'political system'), and the third certainly in only three (competitive, participatory, state-sponsored). Most strikingly of all: *none* of the 'decisive' dimensions, it is to be noted, figures obviously in one elaboration of the five 'models of trade union movements' (opposition).

This confusion of classificatory criteria concerning 'models of trade union movements' (in a study concerned with 'National Trade Union Movements') is compounded when in the case of two specified countries the typological issue is resolved with reference to a single trade union organization which does not stand alone in its national context, but confronts two or more equivalent bodies. Thus the Italian trade union movement is identified with *two* categories, as the table shows—one on the basis of a single nominated body (CGIL: the Italian General Confederation of Labour), and the other, presumably, on the basis of the CGIL's two competitors (the Confederation of Workers' Unions and the Italian Workers' Union).[19] The French trade union movement, in contrast, is allocated to a single category, but explicitly on the basis of only one (CGT: General Confederation of Labour) of its four trade union confederations, even though the other three would seem to be in the same position as those given a second category in the Italian case.

[19] See ibid. 220.

BIBLIOGRAPHY TO PART I

ABELL, AARON I. (ed.), *American Catholic Thought on Social Questions*, New York: Bobs-Merrill, 1968.

ALLEN, V. L., *Militant Trade Unionism*, London: Merlin Press, 1966.

—— *The Sociology of Industrial Relations*, London: Longmans, 1971.

ANDERSON, THORNTON, *Masters of Russian Marxism*, New York: Appleton-Century-Crofts, 1963.

ARONSON, ROBERT L. (ed.), 'The Theory of the Labor Movement Reconsidered: A Symposium in Honor of Selig Perlman', *Industrial and Labor Relations Review*, 13/3, Apr. 1960, pp. 334–97.

ATHERTON, WALLACE N., *Theory of Union Bargaining Goals*, Princeton: Princeton University Press, 1973.

BAIN, GEORGE S., and PRICE, ROBERT, *Profiles of Union Growth*, Oxford: Blackwell, 1980.

BAKKE E. W., KERR, C., and ANROD, C. W., *Unions, Management and the Public*, New York: Harcourt, Brace, 1967.

BAKUNIN, MICHAEL, *Marxism, Freedom and the State*, London: Freedom Press, 1950.

BANKS, J. A., *Trade Unionism*, London: Collier-Macmillan, 1974.

BARBASH, JACK, 'Consumption Values of Trade Unions', *Journal of Economic Issues*, 7, June 1973, pp. 289–301.

BARNARD, CHESTER I., *The Functions of the Executive*, Cambridge, Mass.: Harvard University Press, 1938.

BARNES, SAMUEL H., 'The Evolution of Christian Trade Unionism in Quebec', *Industrial and Labor Relations Review*, 12/4, July 1959, pp. 568–91.

BAUDER, RUSSELL, 'Three Interpretations of the American Labor Movement', *Social Forces*, 22, Dec. 1943, pp. 215–24.

BELING, WILLARD A. (ed.), *The Role of Labor in African Nation-Building*, New York: Praegar, 1968.

BLACKBURN, ROBIN, and COCKBURN, A. (eds.), *The Incompatibles: Trade Union Militancy and the Consensus*, Harmondsworth: Penguin, 1967.

BLAU, PETER M., and SCOTT, W. R., *Formal Organizations: A Comparative Approach*, London: Routledge, 1963.

BOGGS, CARL, *Gramsci's Marxism*, London: Pluto Press, 1976.

BOOTHBY, ROBERT, LODER, J., MACMILLAN, H., and STANLEY, O., *Industry and the State: A Conservative View*, London: Macmillan, 1927.

BRADY, ROBERT A., *The Spirit and Structure of German Fascism*, London: Gollancz, 1937.

BRICIANER, SERGE, *Pannekoek and the Workers' Councils*, Saint Louis: Telos Press, 1978.

Brissenden, Paul F., *The I.W.W.: A Study of American Syndicalism*, 2nd edn., New York: Russell, 1957.

Carlson, Peter, *Roughneck: The Life and Times of Big Bill Haywood*, New York: Norton, 1983.

Carlyle, Thomas, *Past and Present*, London: Ward Lock, 1911.

—— *Scottish and Other Miscellanies*, London: Dent, 1915.

Carr, E. H., *The Bolshevik Revolution 1917–1923*, vol. ii, London: Macmillan, 1952.

Catholic Encyclopedia, New, vol. viii ('Labor Movement'), New York: McGraw-Hill, 1967.

Cecil, Lord Hugh, *Conservatism*, London: Williams and Norgate, 1912.

Child, John, Loveridge, R., and Warner, M., 'Toward an Organizational Study of Trade Unions', *Sociology*, 7, 1973, pp. 71–91.

Clarke, Peter B., and Wilson, J. Q., 'Incentive Systems: A Theory of Organizations', *Administrative Science Quarterly*, 6/2, Sept. 1961, pp. 129–66.

Clarke, T., and Clements, L. (eds.), *Trade Unions under Capitalism*, London: Fontana, 1977.

Clegg, H. A., *The Changing System of Industrial Relations in Great Britain*, Oxford: Blackwell, 1979.

—— *Industrial Democracy and Nationalization*, Oxford: Blackwell, 1951.

—— *A New Approach to Industrial Democracy*, Oxford: Blackwell, 1960.

—— 'Pluralism in Industrial Relations', *British Journal of Industrial Relations*, 13/3, Nov. 1975, pp. 309–16.

Cole, G. D. H., *Chaos and Order in Industry*, London: Methuen, 1920.

—— *Guild Socialism Re-stated*, London: Parsons, 1920.

—— *A History of Socialist Thought*, vol. i, London: Macmillan, 1959.

—— *Self-Government in Industry*, London: Bell, 1918.

—— *The World of Labour*, London: Bell, 1913.

Collins, Henry, and Abramsky, C., *Karl Marx and The British Labour Movement*, London: Macmillan, 1965.

Conquest, Robert, *Industrial Workers in the USSR*, London: Bodley Head, 1967.

Crouch, Colin, *Trade Unions: The Logic of Collective Action*, Glasgow: Fontana, 1982.

Dabscheck, Braham, 'Of Mountains and Routes Over Them: A Survey of Theories of Industrial Relations', *Journal of Industrial Relations*, 25/4, Dec. 1983, pp. 485–506.

De Brizzi, John A., *Ideology and The Rise of Labor Theory in America*, Westport, Conn.: Greenwood Press, 1983.

De Leon, Daniel, *As To Politics*, 2nd edn., New York: Socialist Labor Party, 1915.

De Leon, Daniel, *The Burning Question of Trades Unionism*, New York: Labor News, 1977.
—— *Industrial Unionism: Selected Editorials*, 5th edn., New York: Labor News, 1963.
—— *Socialist Reconstruction of Society: The Industrial Vote*, New York: Labor News, 1977.
Derber, Milton, *The American Idea of Industrial Democracy 1865–1965*, Urbana: University of Illinois Press, 1970.
Deutscher, Isaac, *The Prophet Armed: Trotsky 1879–1921*, London: Oxford University Press, 1970.
Dolgoff, Sam, *Bakunin on Anarchy*, London: Allen and Unwin, 1973.
Dunlop, John T., *Wage Determination under Trade Unions*, New York: Kelley, 1950.
Ely, Richard T., *Ground under Our Feet*, New York: Macmillan, 1938.
—— *The Labor Movement in America*, London: Heinemann, 1890.
—— and Wicker, G. R., *Elementary Principles of Economics*, London: Macmillan, 1925.
Engels, F., *The Condition of the Working Class in England*, Moscow: Progress Publishers, 1973.
Estey, James A., *Revolutionary Syndicalism: An Exposition and a Criticism*, London: King, 1913.
Etzioni, Amitai, *A Comparative Analysis of Complex Organizations*, 2nd edn., Glencoe: Free Press, 1975.
—— *Modern Organizations*, Englewood Cliffs, NJ: Prentice-Hall, 1964.
—— (ed.), *A Sociological Reader on Complex Organizations*, 2nd edn., London: Holt, Rinehart and Winston, 1970.
Flanders, Allan, *The Fawley Productivity Agreements*, London: Faber, 1964.
—— *Management and Unions: The Theory and Reform of Industrial Relations*, London: Faber, 1975.
Fogarty, Michael P., *Christian Democracy in Western Europe 1820–1953*, London: Routledge, 1957.
Freeman, Richard B., and Medoff, J. L., *What Do Unions Do?*, New York: Basic Books, 1984.
Fremantle, Anne (ed.), *The Papal Encyclicals in Their Historical Context*, New York: Mentor-Omega, 1963.
Friedland, William H., and Rosberg, C. G. (eds.), *African Socialism*, Stanford: Hoover Institution, 1964.
Friedman, Milton, *Capitalism and Freedom*, Chicago: University of Chicago Press, 1962.
—— and Friedman, Rose, *Free to Choose*, London: Secker and Warburg, 1980.

GILSON, ETIENNE (ed.), *The Church Speaks to the Modern World: The Social Teachings of Leo XIII*, New York: Doubleday, 1954.

GLASS, S. T., *The Responsible Society: The Ideas of Guild Socialism*, London: Longmans, 1966.

GOLDSWORTHY, DAVID, *Tom Mboya: The Man Kenya Wanted to Forget*, London: Heinemann, 1982.

GOODRICH, CARTER L., *The Frontier of Control*, New York: Harcourt, Brace, and Howe, 1920.

GRAMSCI, ANTONIO, *Selections from Political Writings 1910–1920*, London: Lawrence and Wishart, 1977.

—— *Selections from Political Writings 1921–1926*, London: Lawrence and Wishart, 1978.

—— *Selections from the Prison Notebooks of Antonio Gramsci* (ed. Quintin Hoare and G. N. Smith), London: Lawrence and Wishart, 1971.

GREENSLADE, S. L., *The Church and The Social Order*, London: SCM Press, 1948.

GUÉRIN, DANIEL, *Anarchism: From Theory to Practice*, New York: Monthly Review Press, 1970

GULICK, L., and BERS, M., 'Insight and Illusion in Perlman's Theory of the Labor Movement', *Industrial and Labor Relations Review*, 6/4, July 1953, pp. 510–31.

HAILSHAM, Lord (Quintin Hogg), *The Case for Conservatism*, West Drayton: Penguin, 1947.

—— *The Conservative Case*, Harmondsworth: Penguin, 1959.

—— *The Dilemma of Democracy*, London: Collins, 1978.

HALEY, J. H., *Syndicalism*, London: Jack, n.d.

HAMMOND, THOMAS T., *Lenin on Trade Unions and Revolution 1893–1917*, New York: Columbia University Press, 1957.

HAYEK, F. A., *The Constitution of Liberty*, London: Routledge, 1960.

—— *1980s Unemployment and The Unions*, London: Institute of Economic Affairs, 1980.

HAYES, PAUL M., *Fascism*, London: Allen and Unwin, 1973.

HEIMANN, EDUARD, *Reason and Faith in Modern Society: Liberalism, Marxism and Democracy*, Middletown, Conn.: Wesleyan Union Press, 1961.

HELLER, Joseph, *Good as Gold*, New York: Pocket Books, 1980.

HILL, JOHNSON D., and STUERMANN, W. E., *Organized Labor, A Philosophical Perspective*, New York: Exposition Press, 1961.

HILLQUIT, MORRIS, and RYAN, JOHN A., *Socialism: Promise or Menace?*, New York: Macmillan, 1917.

HILTER, ADOLF, *My Struggle*, London: Paternoster, 1933.

HOBSON, S. G., *Guild Principles in War and Peace*, London: Bell, *c.*1916.

Hobson, S. G., *National Guilds and the State*, London: Bell, 1920.
—— *National Guilds: An Inquiry into the Wage System*, London: Bell, 1914.
Hodges, Donald C., 'The Rise and Fall of Militant Trade Unionism', *American Journal of Economics and Sociology*, 20, Oct. 1961, pp. 438–96.
Hogg, Quintin, *see* Hailsham.
Holton, Bob, *British Syndicalism 1910–14*, London: Pluto Press, 1976.
Horowitz, Irving Louis, *Radicalism and The Revolt Against Reason: The Social Theories of Georges Sorel*, London: Routledge, 1961.
Howard, W. A., 'Australian Trade Unions in the Context of Union Theory', *Journal of Industrial Relations*, 13/3, Sept. 1977, pp. 255–73.
Howell, George, *Conflicts of Capital and Labour*, 2nd edn. London: Macmillan, 1890.
Hoxie, Robert F., *Trade Unionism in the United States*, New York: Russell, 1966.
Hyman, Richard, *Marxism and the Sociology of Trade Unionism*, London: Pluto Press, 1971.
—— *The Workers' Union*, London: Oxford University Press, 1971.
—— and Fryer, Bob, 'Trade Unions', in John B. McKinlay (ed.), *Processing People: Cases in Organizational Behaviour*, London: Holt, Rinehart, 1975.
International Labour Office, *Freedom of Association: An International Survey*, Geneva, 1975.
James, Walter, *The Christian in Politics*, London: Oxford University Press, 1962.
John Paul ii, Pope, *Laborem Exercens: On Human Work*, London: Catholic Truth Society, 1981.
Joll, James, *The Anarchists*, 2nd edn., London: Methuen, 1979.
Jones, Peter d'A., *The Christian Socialist Revival 1877–1914: Religion, Class and Social Conscience in Late-Victorian England*, Princeton: Princeton University Press, 1968.
Kabos, E., and Zsilak, A., *Studies on the History of the Hungarian Trade-Union Movement*, Budapest: Akadémiai Kiado, 1977.
Kamaliza, Michael, 'Tanganyika's View of Labour's Role', *East African Journal*, Nov. 1964, pp. 9–16.
Karson, Marc, *American Labor Unions and Politics 1900–1918*, Carbondale: Southern Illinois University Press, 1958.
Kassalow, Everett M., *Trade Unions and Industrial Relations: An International Comparison*, New York: Random House, 1969.
Kaufmann, M., *Christian Socialism*, London: Kegan Paul, 1888.
Kernig, C. D. (ed.), *Marxism, Communism and Western Society: A Comparative Encyclopaedia*, vol. viii, New York: Herder and Herder, 1973.

KERR, CLARK, DUNLOP, J. T., HARBISON, F. H., and MYERS, C. A., *Industrialism and Industrial Man*, 2nd. edn., New York: Oxford University Press, 1964.

KORNBLUH, JOYCE L. (ed.), *Rebel Voices: An I.W.W. Anthology*, Ann Arbor: University of Michigan Press, 1964.

LAPIDES, KENNETH (ed.), *Marx and Engels on the Trade Unions*, New York: Praegar, 1987.

LARSON, SIMEON, and NISSEN, B. (eds.), *Theories of the Labor Movement*, Detroit: Wayne State University Press, 1987.

LENIN, V. I., *Collected Works*, London: Lawrence and Wishart, 1960–80.

LESTER, R. A., and SHISTER, J. (eds.), *Insights into Labor Issues*, New York: Macmillan, 1948.

LIPSET, S. M., 'Trade Unions and Social Structure', *Industrial Relations*, 1/1, Oct. 1961, pp. 75–89; and 1/2, Feb. 1962, pp. 89–110.

LOZOVSKY, A. (ed.), *Handbook on the Soviet Trade Unions for Workers' Delegations*, Moscow: Cooperative Publishing Society, 1937.

—— *Marx and the Trade Unions*, London: Lawrence and Wishart, 1935.

LUBEMBE, CLEMENT K., 'Trade Unions and Nation Building', *East African Journal*, Apr. 1964, pp. 19–22.

LUXEMBURG, ROSA, *Selected Political Writings of Rosa Luxemburg* (ed. Dick Howard), New York: Monthly Review Press, 1971.

MCCARTHY, W. E. J. (ed.), *Trade Unions*, Harmondsworth: Penguin, 1972.

—— (ed.), *Trade Unions*, 2nd edn., Harmondsworth: Penguin, 1985.

MACDONALD, J. RAMSAY, *Syndicalism*, London: Constable, 1912.

MCKEE, DON K., 'Daniel De Leon: A Reappraisal', *Labor History*, 1, Fall 1960, pp. 264–97.

MCLELLAN, DAVID, *Karl Marx: Selected Writings*, Oxford: Oxford University Press, 1977.

—— *Marxism after Marx*, London: Macmillan, 1979.

—— *The Thought of Karl Marx: An Introduction*, London: Macmillan, 1971.

MACMILLAN, HAROLD, *The Middle Way*, Wakefield: E. P. Publishing, 1978.

—— *Reconstruction: A Plea for a National Policy*, London: Macmillan, 1933.

NCNEAL, ROBERT H. (general ed.), *Resolutions and Decisions of the Communist Party of The Soviet Union*, 5 vols., Toronto: University of Toronto Press, 1974–82.

MARCH, JAMES G., and SIMON, H. A., *Organizations*, New York: Wiley, 1965.

MARITAIN, JACQUES, *Scholasticism and Politics*, 2nd edn., London: Bles, 1945.

MARTIN, DONALD L., *An Ownership Theory of the Trade Union*, Berkeley: University of California Press, 1980.

MARTIN, ROSS M., *TUC: The Growth of a Pressure Group 1868–1976*, Oxford: Clarendon Press, 1980.

MBOYA, TOM, *The Challenge of Nationhood*, New York: Praegar, 1970.

—— *Freedom and After*, London: Deutsch, 1963.

MEADOWS, MARTIN, 'A Managerial Theory of Unionism', *American Journal of Economics and Sociology*, 25, Apr. 1966, pp. 127–40.

MILLEN, BRUCE H., *The Political Role of Labor in Developing Countries*, Honolulu: East–West Centre Press, 1963.

MILLER, DAVID, *Anarchism*, London: Dent, 1984.

MITCHELL, DANIEL J. B., 'Union Wage Policies: the Ross–Dunlop Debate Re-opened', *Industrial Relations*, 11/2, Feb. 1972, pp. 46–61.

MOODY, JOSEPH N., 'Leo XIII and the Social Crisis', in E. T. Gargan (ed.), *Leo XIII and the Modern World*, New York: Sheed and Ward, 1961.

MOON, PARKER THOMAS, *The Labour Problem and the Social Catholic Movement in France*, New York: Macmillan, 1921.

MOORE, WILBERT E., 'Notes for a General Theory of Labor Organizations', *Industrial and Labor Relations Review*, 13/3, Apr. 1960, pp. 387–97.

MUSSOLINI, BENITO, *Fascism: Doctrine and Institutions*, New York: Fertig, 1968.

—— *My Autobiography*, London: Paternoster, 1928.

NETTL, J. P., *Rosa Luxemburg*, 2 vols., London: Oxford University Press, 1966.

NEUMANN, FRANZ, *Behemoth: The Structure and Practice of National Socialism*, London: Gollancz, 1943.

OAKESHOTT, MICHAEL (ed.), *Social and Political Doctrines of Contemporary Europe*, New York: Cambridge University Press, 1950.

OLSON, MANCUR, *The Logic of Collective Action*, Cambridge, Mass.: Harvard University Press, 1971.

ONUOHA, Father BEDE, *The Elements of African Socialism*, London: Deutsch, 1965.

ORAGE, A. R. (ed.), *National Guilds: An Inquiry into the Wage System*, London: Bell, 1914.

O'SULLIVAN, NOEL, *Conservatism*, London: Dent, 1976.

PANNEKOEK, ANTON, *Lenin as Philosopher*, New York: New Essays, 1948.

PATAUD, ÉMILE, and POUGET, E., *Syndicalism and the Cooperative Commonwealth*, Oxford: New International Publishing, 1913.

PENTY, A. J., *Restoration of the Guild System*, London: Swann Sonnenschein, 1906.

PERLMAN, MARK, 'Labor Movement Theories: Past, Present and Future', *Industrial and Labor Relations Review*, 13/3, Apr. 1960, pp. 338–48.

—— *Labor Union Theories in America*, Evanston: Row, Peterson, 1958.

—— 'Theories of the Labor Movement', *International Encyclopaedia of the Social Sciences*, vol. viii, New York: Macmillan, 1968.

PERLMAN, SELIG, 'The Principle of Collective Bargaining', *Annals of the American Academy of Political and Social Science*, 184, 1936, pp. 154–60.

—— *Selig Perlman's Lectures on Capitalism and Socialism*, Madison: University of Wisconsin Press, 1976.

—— *A Theory of the Labor Movement*, New York: Kelley, 1949.

PERÓN, EVA DUARTE, *Evita by Evita*, London: Proteus, 1978.

PERÓN, JUAN D., *Perón Expounds His Doctrine*, Buenos Aires: n.p., 1948.

—— *Peronist Doctrine*, Buenos Aires: n.p., 1952.

—— *The Voice of Perón*, Buenos Aires: n.p., 1950.

PETRO, SYLVESTER, *The Labor Policy of the Free Society*, New York: Ronald Press, 1957.

POOLE, MICHAEL, *Theories of Trade Unionism: A Sociology of Industrial Relations*, London: Routledge, 1981.

RAUSCHENBUSCH, WALTER, *Christianity and the Social Crisis*, New York: Harper and Row, 1964.

RAVEN, CHARLES E., *Christian Socialism 1848–1854*, London: Cass, 1968.

RENSHAW, P., *The Wobblies: The Story of Syndicalism in the United States*, London: Eyre and Spottiswoode, 1967.

RIDLEY, F. F., *Revolutionary Syndicalism in France: The Direct Action of its Time*, Cambridge: Cambridge University Press, 1970.

ROSE, PETER I. (ed.), *The Study of Society: An Integrated Anthology*, New York: Random House, 1967.

ROSS, ARTHUR M., *Trade Union Wage Policy*, Berkeley: University of California Press, 1948.

RUBEL, MAXIMILIEN, *Marx: Life and Works*, London: Macmillan, 1980.

RUSSELL, BERTRAND, *Roads to Freedom: Socialism, Anarchism and Syndicalism*, London: Allen and Unwin, 1918.

RYAN, JOHN AUGUSTINE, *Declining Liberty and Other Papers*, New York: Macmillan, 1927.

—— 'Moral Aspects of Labor Unions', in *Catholic Encyclopedia*, vol. viii, New York: Encyclopedia Press, 1910, pp. 724–8.

—— and HILLQUIT, M., *Socialism: Promise or Menace?*, New York: Macmillan, 1914.

SCHMIDMAN, JOHN, *Unions in Post-industrial Society*, University Park, Penn.: Pennsylvania State University Press, 1979.

SCHMIDT, CARL T., *The Corporate State in Action: Italy Under Fascism*, London: Gollancz, 1939.

SCHOENBAUM, DAVID, *Hitler's Social Revolution*, London: Weidenfeld and Nicolson, 1967.

SCHUMPETER, JOSEPH A., *Capitalism, Socialism and Democracy*, London: Allen and Unwin, 1954.

SCOTT, JOHN W., *Syndicalism and Philosophical Realism*, London: Black, 1919.

SCRUTON, ROGER, *The Meaning of Conservatism*, Harmondsworth: Penguin, 1980.
SHAW, GEORGE BERNARD, *et al.*, *Fabian Essays*, 6th edn., London: Allen and Unwin, 1962.
SIEGEL, ABRAHAM J., 'The Extended Meaning and Diminished Relevance of "Job Conscious" Unionism', in Industrial Relations Research Association, *Proceedings, 1965*, Madison, 1966, pp. 166–82.
SILVERMAN, DAVID, *The Theory of Organisations*, London: Heinemann, 1970.
SIMON, HERBERT, *Administrative Behaviour: A Study of Decision Making Processes in Administrative Organizations*, New York: Macmillan, 1976.
SMART, D. A. (ed.), *Pannekoek and Gorter's Marxism*, London: Pluto Press, 1978.
SMITH, ADAM, *An Inquiry into the Nature and Causes of the Wealth of Nations*, 4th edn., Edinburgh: Black, 1850.
SMITH, PAUL (ed.), *Lord Salisbury on Politics: A Selection from His Articles in the Quarterly Review, 1860–1883*, Cambridge: Cambridge University Press, 1972.
SOREL, GEORGES, *Reflections on Violence*, London: Collier-Macmillan, 1960.
SPARGO, JOHN, *Syndicalism, Industrial Unionism and Socialism*, New York: Huebsch, 1913.
SRIVASTVA, SURESH CHANDRA, 'Trade Unions' Participation in the Planned Economy of India', *Journal of Industrial Relations*, 12/2, July 1970, pp. 238–52.
STALIN, J. V., *On the Opposition (1921–27)*, Peking: Foreign Languages Press, 1974.
—— *Problems of Leninism*, Moscow: Foreign Languages Publishing House, 1953.
STANLEY, JOHN, *The Sociology of Virtue: the Political and Social Theories of Georges Sorel*, Berkeley: University of California Press, 1981.
STEARNS, PETER N., *Revolutionary Syndicalism and French Labor: A Cause without Rebels*, Brunswick, NJ: Rutgers University Press, 1971.
STUDDERT-KENNEDY, GERALD, *Dog-collar Democracy: The Industrial Christian Fellowship, 1919–1929*, London: Macmillan, 1982.
STURMTHAL, ADOLF F., 'Comments on Selig Perlman's *A Theory of the Labor Movement*', *Industrial and Labor Relations Review*, 4, 1951, pp. 483–96.
—— and SCOVILLE, J. C. (eds.), *The International Labor Movement in Transition*, Urbana: University of Illinois Press, 1973.
SUCHTING, W. A., *Marx: An Introduction*, Brighton: Harvester Press, 1983.

SUFRIN, S. C., *Unions in Emerging Societies: Frustration and Politics*, Syracuse: Syracuse University Press, 1964.

TAFT, PHILIP, 'Reflections on Selig Perlman as a Teacher and Writer', *Industrial and Labor Relations Review*, 29/2, Jan. 1976, pp. 249–57.

—— 'Theories of the Labor Movement', in G. W. Brookes *et al.* (eds.), *Interpreting the Labor Movement*, Madison: Industrial Relations Research Association, 1952, pp. 1–38.

TANNENBAUM, FRANK, *The Labor Movement: Its Conservative Functions and Social Consequences*, New York: Putnams, 1921.

—— 'The Social Functions of Trade Unionism', *Political Science Quarterly*, 62, June 1947, pp. 161–94.

—— *The True Society: A Philosophy of Labour*, London: Cape, 1964 (published as *A Philosophy of Labor*, New York: Knopf, 1952).

TAWNEY, R. H., *The Acquisitive Society*, London: Fontana, 1961.

—— *The Attack and Other Papers*, London: Allen and Unwin, 1953.

—— *The British Labor Movement*, New York: Greenwood Press, 1968.

—— *Equality*, London: Allen and Unwin, 1931.

—— *The Radical Tradition: Twelve Essays on Politics, Education and Literature* (ed. Rita Hinden), London: Allen and Unwin, 1964.

TEMPLE, WILLIAM, *Christianity and Social Order*, London: Shepheard-Walwyn, 1976.

THOMPSON, J. D., *Organizations in Action*, New York; McGraw-Hill, 1967.

TREACY, GERALD C., and GIBBONS, W. J. (eds.), *Seven Great Encyclicals*, New York: Paulist Press, 1963.

TROELTSCH, ERNST, *The Social Teachings of the Christian Churches*, 2 vols., London: Allen and Unwin, 1950.

TUCKER, ROBERT C. (ed.), *The Marx–Engels Reader*, 2nd edn., New York: Norton, 1978.

UTLEY, T. E., *Essays in Conservatism*, London: Conservative Political Centre, 1949.

VIERECK, PETER, *Conservatism: from John Adams to Churchill*, Princeton: Nostrand, 1956.

WEBB, BEATRICE (ed. N. and J. MACKENZIE), *The Diary of Beatrice Webb, 1924–43*, vol. iv, London: Virago and LSE, 1985.

WEBB, SIDNEY and BEATRICE, *History of Trade Unionism*, London: Longmans, 1894.

—— *History of Trade Unionism*, 2nd edn., London: Longmans, 1920.

—— *Industrial Democracy*, London: Longmans, 1902.

—— *Soviet Communism: A New Civilization?*, London: Longmans, Green, 1935.

—— *Soviet Communism: A New Civilization*, 2nd edn., London: Longmans, 1937.

WHITE, R. J. (ed.), *The Political Thought of Samuel Taylor Coleridge*, London: Cape, 1938.

WOODCOCK, GEORGE, *Anarchism*, Harmondsworth: Penguin, 1963.

—— (ed.), *The Anarchist Reader*, Glasgow: Fontana, 1977.

WOOTTON, GRAHAM, *Workers, Unions and The State*, London: Routledge, 1966.

BIBLIOGRAPHY TO PART II

There is a huge literature, in English, relating to trade unions in different countries of the world. Often, especially in the case of less prominent countries, it involves works which are not primarily concerned with trade unionism. Apart from this, much of the material is of largely historical interest. A great deal of it, too, is of dubious quality, being either highly formal in approach, simplistic, crudely polemical, or a combination of these things.

The selection listed below is confined to references which are quoted or cited in the text, or were otherwise found to be of particular relevance to the study. So far as possible, the emphasis is on good *contemporary* references, but the notion of 'contemporary' has had to be stretched in many cases—on grounds of either quality or simple availability, or both.

Items are grouped in two sections. The first section is concerned with general references and multi-country studies; the second with works confined to a single country. Much of the material in the text on particular countries is drawn from multi-country studies. It was not practicable, however, to list in the bibliography the countries covered in the case of such studies.

The single-country references are ordered by country, alphabetically, in relation to the twenty-seven countries to which particular attention has been paid in the text. A number of single-country references relating to some of the other countries mentioned in the text are listed at the end.

GENERAL AND COMPARATIVE (MULTI-COUNTRY) REFERENCES

AARON, BENJAMIN, and WEDDERBURN, K. W. (eds.), *Industrial Conflict: A Comparative Legal Survey*, New York: Crane, Russak, 1972.

ABELL, AARON I. (ed.), *American Catholic Thought on Social Questions*, New York: Bobs-Merrill, 1968.

ALBA, VICTOR, *Politics and The Labor Movement in Latin America*, Stanford: Stanford University Press, 1968.

ALMOND, GABRIEL A., 'Corporatism, Pluralism, and Professional Memory', *World Politics*, 35/2, Jan. 1983, pp. 251–60.

ANANABA, WOGU, *The Trade Union Movement in Africa: Promise and Performance*, London: Hurst, 1979.

'Asia's Unions', *Far Eastern Economic Review*, 132/14, 3 Apr. 1986, pp. 43–67.

BAIN, GEORGE S., and PRICE, ROBERT, *Profiles of Union Growth: A Comparative Statistical Portrait of Eight Countries*, Oxford: Blackwell, 1980.

BALL, ALAN R., and MILLARD, F., *Pressure Politics in Industrial Societies: A Comparative Introduction*, London: Macmillan, 1986.

BAMBER, GREG J., and LANSBURY, R. D. (eds.), *International and Comparative Industrial Relations: A Study of Developed Market Economies*, Sydney: Allen and Unwin, 1987.

BARBASH, JACK, *Trade Unions and National Economic Policy*, Baltimore: Johns Hopkins Press, 1972.

BARKIN, SOLOMON, *et al.* (eds.), *International Labor*, New York: Harper and Row, 1967.

BEAN, R., *Comparative Industrial Relations: An Introduction to Cross-National Perspectives*, London: Croom Helm, 1985.

BELING, WILLARD A. (ed.), *The Role of Labor in African Nation-Building*, New York: Praegar, 1968.

BEYME, KLAUS VON, *Challenge to Power: Trade Unions and Industrial Relations in Capitalist Countries*, London: Sage, 1980.

—— *Political Parties in Western Democracies*, Aldershot: Gower, 1985.

BLANPAIN, R. (ed.), *Comparative Labour Law and Industrial Relations*, 2nd edn., Deventer: Kluwer Law, 1985.

BLAU, PETER M., and SCOTT, W. R., *Formal Organizations: A Comparative Approach*, London: Routledge, 1966.

BLUM, A. A. (ed.), *International Handbook of Industrial Relations: Contemporary Developments and Research*, Westport, Conn.: Greenwood Press, 1981.

Catholic Encyclopedia, New vol. viii ('Labor Movement'), New York: McGraw-Hill, 1967.

CAWSON, ALAN, *Corporatism and Political Theory*, Oxford: Blackwell, 1986.

CELLA, GIAN P., and TREU, T., 'National Trade Union Movements', in R. Blanpain (ed.), *Comparative Labour Law and Industrial Relations*, 2nd edn., Deventer: Kluwer Law, 1985.

CLARK, JON, HARTMANN, H., LAU, C., and WINCHESTER, D., *Trade Unions, National Politics and Economic Management: A Comparative Study of the TUC and the DGB*, London: Anglo-German Foundation for the Study of Industrial Society, 1980.

CLEGG, H. A., *Trade Unionism under Collective Bargaining: A Theory Based on Comparisons of Six Countries*, Oxford: Blackwell, 1976.

COLEMAN, JAMES S., and ROSBERG, C. E. (eds.), *Political Parties and National Integration in Tropical Africa*, Berkeley: University of California Press, 1964.

COOMBES, DAVID, 'Trade Unions and Political Parties in Britain, France, Italy and West Germany', *Government and Opposition*, 13/4, Autumn 1978, pp. 485–95.

CORDOVA, EFREN (ed.), *Industrial Relations in Latin America*, New York: Praegar, 1984.

DAALDER, HANS, and MAIR, P. (eds.), *Western European Party Systems*, London: Sage, 1983.

DAMACHI, UKANDI G., SEIBEL, H. DIETER, and TRACHTMAN, LESTER (eds.), *Industrial Relations in Africa*, New York: St Martin's Press, 1979.

DAVIES, IOAN, *African Trade Unions*, Harmondsworth: Penguin, 1966.

DAVIS, CHARLES L., and COLEMAN, K. M., 'Labor and the State: Union Incorporation and Working-class Politicization in Latin America', *Comparative Political Studies*, 18/4, Jan. 1986, pp. 395–418.

DUNLOP, JOHN T., *Industrial Relations Systems*, New York: Henry Holt, 1958.

—— and GALENSON, W. (eds.), *Labor in the Twentieth Century*, New York: Academic Press, 1978.

DUVERGER, MAURICE, *Political Parties: Their Organization and Activity in the Modern State*, London: Methuen, 1954.

ELVANDER, N., 'The Role of the State in Settlement of Labour Disputes in the Nordic Countries: A Comparative Analysis', *European Journal of Political Research*, 2, 1974, pp. 363–83.

ETZIONI, AMITAI, *A Comparative Analysis of Complex Organizations*, New York: Free Press, 1961.

FOGARTY, MICHAEL P., *Christian Democracy in Western Europe 1820–1953*, London: Routledge, 1957.

FRANK, ANDRE GUNDER, *Capitalism and Underdevelopment in Latin America: Historical Studies of Chile and Brazil*, Harmondsworth: Penguin, 1971.

FRIEDLAND, WILLIAM H., *Unions and Industrial Relations in Underdeveloped Countries*, Bulletin 47, New York State School of Industrial and Labor Relations, Cornell University, Jan. 1963.

FRIEDMAN, MILTON and ROSE, *Free to Choose*, London: Secker and Warburg, 1980.

FRIEDRICH, CARL J., and BRZEZINSKI, Z. K., *Totalitarian Dictatorship and Autocracy*, 2nd edn., New York: Praegar, 1966.

GALENSON, WALTER (ed.), *Comparative Labor Movements*, New York: Prentice-Hall, 1952.

—— (ed.), *Labor and Economic Development*, New York: Wiley, 1959.

—— (ed.), *Labor in Developing Economies*, Berkeley: University of California Press, 1962.

GALLIE, DUNCAN, *Social Inequality and Class Radicalism in France and Britain*, Cambridge: Cambridge University Press, 1983.

GHOSH, SUBRATESH, *Trade Unionism in the Underdeveloped Countries*, Calcutta: Bookland, 1960.

HAYWARD, JACK (ed.), *Trade Unions and Politics in Western Europe*, London: Cass, 1980.

HENLEY, JOHN, 'The Management of Labour Relations in Industrialising Market Economies', *Industrial Relations Journal*, 10, 1980, pp. 41–53.

HOLMES, LESLIE, *Politics in the Communist World*, Oxford: Clarendon Press, 1986.

HUNTINGTON, SAMUEL P., and MOORE, C. H. (eds.), *Authoritarian Politics in Modern Society: The Dynamics of Established One-Party Systems*, New York: Basic Books, 1970.

INDUSTRIAL DEMOCRACY IN EUROPE. (International Research Group), *European Industrial Relations*, Oxford: Clarendon Press, 1981.

INTERNATIONAL LABOUR OFFICE, *Collective Bargaining in Industrialised Market Economies*, Geneva, 1973.

—— *Freedom of Association: An International Survey*, Geneva, 1975.

—— *Year Book of Labour Statistics*, Geneva.

JACKSON, MICHAEL P., *Industrial Relations*, London: Croom Helm, 1977.

JACOBS, ERIC, *European Trade Unionism*, London: Croom Helm, 1973.

JURIS, HERVEY, *et al.* (eds.), *Industrial Relations in a Decade of Economic Change*, Madison: Industrial Relations Research Association, 1985.

KASSALOW, EVERETT M., 'The Comparative Labour Field', *Bulletin of the International Institute for Labour Studies*, 5, July 1968, pp. 92–107.

—— (ed.), *National Labor Movements in the Post-war World*, Evanston, Illinois: North-Western University Press, 1963.

—— *Trade Unions and Industrial Relations: An International Comparison*, New York: Random House, 1969.

—— and DAMACHI, U. G. (eds.), *The Role of Trade Unions in Developing Societies*, Geneva: International Institute of Labour Studies, 1978.

KATZENSTEIN, PETER J., *Corporatism and Change: Austria, Switzerland, and the Politics of Industry*, Ithaca: Cornell University Press, 1984.

KENDALL, WALTER, *The Labour Movement in Europe*, London: Allen Lane, 1975.

KERNIG, C. D. (ed.), *Marxism, Communism and Western Society: A Comparative Encyclopaedia*, vol. viii ('Trade Unions'), New York: Herder and Herder, 1973.

KERR, CLARK, DUNLOP, J. T., HARBISON, F. H., and MYERS, C. A., *Industrialism and Industrial Man*, 2nd edn., New York: Oxford University Press, 1964.

KOHLER, BEATE, *Political Forces in Spain, Greece and Portugal*, London: Butterworth, 1982.

KORPI, WALTER, and SHALEV, M., 'Strikes, Industrial Relations and Class Conflict in Capitalist Societies', *British Journal of Sociology*, 30/2, June 1979, pp. 164–87.

LAPALOMBARA, JOSEPH, and WEINER, M. (eds.), *Political Parties and Political Development*, Princeton: Princeton University Press, 1966.

LEHMBRUCH, GERHARD, and SCHMITTER, P. C. (eds.), *Patterns of Corporatist Policy-Making*, London: Sage, 1982.

LENIN, V. I., 'What is to be Done?', *Collected Works*, vol. v, pp. 347–527, London: Lawrence and Wishart, 1961.

LORWIN, VAL R., *Labor and Working Conditions in Modern Europe*, New York: Macmillan, 1967.

LOWIT, T., 'The Working Class and Union Structures in Eastern Europe', *British Journal of Industrial Relations*, 20, Mar. 1982, pp. 67–75.

MCBRIDE, STEPHEN, 'Corporatism, Public Policy and the Labour Movement: A Comparative Study, *Political Studies*, 33/3, Sept. 1985, pp. 439–56.

MARKS, GARY, 'Neocorporatism and Incomes Policy in Western Europe and North America', *Comparative Politics*, 18/3, Apr. 1986, pp. 253–78.

MARTIN, ROSS M., 'The Authority of Trade Union Centres: The Australian Council of Trade Unions and the British Trades Union Congress', *Journal of Industrial Relations*, 4/1, Apr. 1962, pp. 1–19.

—— 'Pluralism and the New Corporatism', *Political Studies*, 31/1, Mar. 1983, pp. 86–102.

—— 'The Problem of "Political" Strikes', in W. A. Howard (ed.), *Perspectives on Australian Industrial Relations*, Melbourne: Longman Cheshire, 1984, pp. 68–81.

—— 'Tribesmen into Trade Unionists: The African Experience and the Papua-New Guinea Prospect', *Journal of Industrial Relations*, 11/2, July 1969, pp. 125–72.

MERKL, PETER H. (ed.), *Western European Party Systems*, New York: Free Press, 1980.

MEYNAUD, JEAN, and BEY, A. S., *Trade Unionism in Africa*, London: Methuen, 1967.

MILLEN, BRUCE H., *The Political Role of Labor in Developing Countries*, Honolulu: East–West Centre Press, 1963.

ORGANIZATION FOR ECONOMIC COOPERATION AND DEVELOPMENT, *Collective Bargaining and Government Policies in Ten Countries: Austria, Canada, France, Germany, Italy, Japan, New Zealand, Sweden, United Kingdom, Unites States*, Paris, 1979.

—— *Labour Disputes: A Perspective*, Paris, 1979.

—— *Wage Policies and Collective Bargaining Developments in Finland, Ireland and Norway*, Paris, 1979.

PERLMAN, SELIG, 'The Principle of Collective Bargaining', *Annals of the American Academy of Political and Social Science*, 184, 1936, pp. 154–60.

—— *A Theory of the Labor Movement*, New York: Kelley, 1949.

POOLE, MICHAEL, *Industrial Relations: Origins and Patterns of National Diversity*, London: Routledge, 1986.

PORKET, J. L., 'Industrial Relations and Participation in Management in the Soviet-type Communist System', *British Journal of Industrial Relations*, 16, Mar. 1978, pp. 70–85.

PRAVDA, ALEX, 'Industrial Workers: Patterns of Dissent, Opposition and Accommodation', in R. L. Tokes (ed.), *Opposition in Eastern Europe*, Baltimore: John Hopkins University Press, 1979, pp. 209–62.

—— 'Trade Unions in East European Communist Systems: Towards Corporatism?', *International Political Science Review*, 4/2, 1983, pp. 241–60.

—— and RUBLE, B. A. (eds.), *Trade Unions in Communist States*, Boston: Allen and Unwin, 1986.

RAFFAELE, J. A., *Labor Leadership in Italy and Denmark*, Madison: University of Wisconsin Press, 1962.

RAWSON, D. W., 'The Life-span of Labour Parties', *Political Studies*, 17/3, Sept. 1969, pp. 313–33.

ROBERTS, B. C. (ed.), *Industrial Relations in Europe: The Imperatives of Change*, London: Croom Helm, 1985.

—— *Labour in the Tropical Territories of the Commonwealth*, London: Bell, 1964.

ROSS, ARTHUR M. (ed.), *Industrial Relations and Economic Development*, London: Macmillan, 1966.

RUSSETT, B., *World Handbook of Political and Social Indicators*, New Haven: Harvard University Press, 1964.

SABEL, C. F., and STARK, D., 'Planning, Politics and Shop-floor Power: Hidden Forms of Bargaining in Soviet-Imposed State-Socialist Societies', *Politics and Society*, 4, 1982, pp. 439–75.

SARTORI, GIOVANNI, *Parties and Party Systems: A Framework for Analysis*, vol. i, Cambridge: Cambridge University Press, 1976.

SCHAPIRO, LEONARD, *Political Opposition in One-party States*, London: Macmillan, 1972.

SCHMITTER, PHILIPPE C., and LEHMBRUCH, G. (eds.), *Trends Towards Corporatist Intermediation*, London: Sage, 1979.

SCHREGLE, JOHANNES, *Negotiating Development: Labour Relations in Southern Asia*, Geneva: International Labour Office, 1982.

SETON-WATSON, HUGH, *The Imperialist Revolutionaries: Trends in World Communism in the 1960s and 1970s*, Stanford: Hoover Institution Press, 1978.

SMITH, E. OWEN (ed.), *Trade Unions in the Developed Economies*, London: Croom Helm, 1981.

STURMTHAL, ADOLF F., *Left of Center: European Labor since World War II*, Urbana: University of Illinois Press, 1983.

—— *Workers Councils: A Study of Workplace Organization on Both Sides of the Iron Curtain*, Cambridge, Mass.: Harvard University Press, 1964.

SUFRIN, S. C., *Unions in Emerging Societies: Frustration and Politics*, Syracuse: Syracuse University Press, 1964.

TANNAHILL, R. NEAL, *The Communist Parties of Western Europe: A Comparative Study*, Westport, Conn.: Greenwood Press, 1978.

TORRINGTON, DEREK (ed.) *Comparative Industrial Relations in Europe*, Westport, Conn.: Greenwood Press, 1978.

TRISKA, JAN F., and GATI, C., (eds.), *Blue-collar Workers in Eastern Europe*, London: Allen and Unwin, 1981.

VALENZUELA, J. SAMUEL, and GOODWIN, JEFFREY, *Labor Movements Under Authoritarian Regimes*, Cambridge, Mass.: Center for European Studies, Harvard University, 1983.

WALLERSTEIN, IMMANUEL (ed.), *Labor in the World Social Structure*, Beverley Hills: Sage, 1983.

WILCZYNSKI, JOSEF, *Comparative Industrial Relations*, London: Macmillan, 1983.

WINDMULLER, JOHN P., 'The Authority of National Trade Union Confederations: A Comparative Analysis', in David B. Lipsky (ed.), *Union Power and Public Policy*, Ithaca: State School of Industrial and Labor Relations, Cornell University, 1975.

—— 'Concentration Trends in Union Structure: An International Comparison', *Industrial and Labor Relations Review*, 35/1, Oct. 1981, pp. 43–57.

WORLD BANK, *The World Development Report*, Washington, DC.

SINGLE-COUNTRY REFERENCES

Australia

BEGGS, JOHN J., and CHAPMAN, B. J., 'Australian Strike Activity in an International Context: 1964–85', *Journal of Industrial Relations*, 29/2, June 1987, pp. 137–49.

DABSCHECK, BRAHAM, and NILAND, J., *Industrial Relations in Australia*, Sydney: Allen and Unwin, 1981.

DEERY, S., and PLOWMAN D., *Australian Industrial Relations*, 2nd edn., Sydney: McGraw-Hill, 1985.

FORD, BILL, and PLOWMAN, D. (eds.), *Australian Unions, An Industrial Relations Perspective*, Melbourne: Macmillan, 1983.

HILL, JOHN D., HOWARD, W. A., and LANSBURY, R. D., *Industrial Relations: An Australian Introduction*, Melbourne: Longman Cheshire, 1982.

MARTIN, ROSS M., 'Australian Trade Unions and Political Action', *Parliamentary Affairs*, 20/1, Winter 1966–7, pp. 35–48.

MARTIN, ROSS M., 'Industrial Relations', in D. M. Gibb and A. W. Hannan (eds.), *Debate and Decision: Political Issues in 20th Century Australia*, Melbourne: Heinemann, 1975, pp. 36–64.

—— 'Political Strikes and Public Attitudes in Australia', *Australian Journal of Politics and History*, 31/2, 1985, pp. 269–81.

—— 'Trade Unions and Labour Governments in Australia: A Study of the Relation between Supporting Interests and Party Policy', *Journal of Commonwealth Political Studies*, 2/1, Nov. 1963, pp. 59–78.

—— *Trade Unions in Australia*, 2nd edn., Harmondsworth: Penguin, 1980.

MITCHELL, D., 'ALP Membership and Affiliation', *Politics*, 16/2, Nov. 1981, pp. 273–5.

PARKIN, ANDREW, and WARHURST, J., (eds.), *Machine Politics in the Australian Labor Party*, Sydney: Allen and Unwin, 1983.

RAWSON, D. W., *Unions and Unionists in Australia*, 2nd edn., Sydney: Allen and Unwin, 1986.

WATERS, MALCOLM, *Strikes in Australia*, Sydney: Allen and Unwin, 1982.

YERBURY, D., 'Legal Regulation of Unions in Australia: The Impact of Compulsory Arbitration and Adversary Politics', in W. A. Howard (ed.), *Perspectives on Australian Industrial Relations*, Melbourne: Longman Cheshire, 1984, pp. 82–103.

Austria

BARBASH, JACK, 'Austrian Trade Unions and the Negotiation of National Economic Policy', *British Journal of Industrial Relations*, 9, Nov. 1971, pp. 371–87.

FURSTENBERG, FRIEDRICH, 'Wage-setting in the Austrian System of Social and Economic Partnership', in R. Blandy and J. Niland (eds.), *Alternatives to Arbitration*, Sydney: Allen & Unwin, 1986.

LACKENBACHER, ERNST, 'Austria', in Adolf Sturmthal (ed.), *White collar Trade Unions*, Urbana: University of Illinois Press, 1967.

MARIN, B., 'Organizing Interests by Interest Associations: Associational Prerequisites of Cooperation in Austria', *International Political Science Review*, 4/2, 1983, pp. 197–216.

STEINER, KURT, *Politics in Austria*, Boston: Little, Brown, 1972.

TOMANDL, THEODOR, and FUERBOECK, KARL, *Social Partnership: The Austrian System of Industrial Relations and Social Insurance*, Ithaca, NY: ILR Press, 1986.

Belgium

BLANPAIN, R., 'Recent Trends in Collective Bargaining in Belgium', *International Labour Review*, 123/3, May–June 1984, pp. 319–32.

LORWIN, VAL R., 'Labor Unions and Political Parties in Belgium', *Industrial and Labor Relations Review*, 28/2, Jan. 1975, pp. 243–63.

Canada

ARTHURS, H. W., CARTER, D. D., and GLASBEEK, H. J., *Labour Law and Industrial Relations in Canada*, Toronto: Butterworths, 1981.

BARNES, SAMUEL H., 'The Evolution of Christian Trade Unionism in Quebec', *Industrial and Labor Relations Review*, 12/4, July 1959, pp. 568–81.

CRISPO, JOHN, *The Canadian Industrial Relations System*, Toronto: McGraw-Hill Ryerson, 1978.

HOROWITZ, GAD, *Canadian Labour in Politics*, Toronto: University of Toronto Press, 1968.

JAMIESON, S., *Industrial Relations in Canada*, 2nd edn., Toronto: Macmillan, 1973.

KWAVNICK, DAVID, *Organized Labour and Pressure Politics: The Canadian Labour Congress 1956–68*, Montreal: McGill-Queen's University Press, 1972.

LAXER, ROBERT, *Canada's Unions*, Toronto: James Lorimer, 1976.

MILLER, RICHARD U., and ISBESTER, F. (eds.), *Canadian Labour in Transition*, Scarborough, Ontario: Prentice-Hall, 1971.

China

FLETCHER, MERTON D., *Workers and Commissars: Trade Union Policy in the People's Republic of China*, Bellingham: Western Washington State College, 1974.

HARPER, PAUL, 'Trade Union Cultivation of Workers for Leadership', in John Wilson Lewis (ed.), *The City in Communist China*, Stanford: Stanford University Press, 1971.

—— 'The Party and the Unions in Communist China', *China Quarterly*, 37, Jan.–Mar. 1969, pp. 84–119.

HEARN, J. M., 'W[h]ither the Trade Unions in China?', *Journal of Industrial Relations*, 19/2, 1977, pp. 158–72.

HENLEY, JOHN S., and CHEN, P. K. N., 'A Note on the Appearance, Disappearance and Re-appearance of Dual Functioning Trade Unions in the People's Republic of China', *British Journal of Industrial Relations*, 19/1, Mar. 1981, pp. 87–93.

LANSBURY, RUSSELL, NG SEK-HONG, and MCKERN, B., 'Management at Enterprise Level In China', *Industrial Relations Journal*, 15/4, Winter 1984, pp. 56–63.

LEE LAI TO, 'The All-China Federation of Trade Unions in the Cultural Revolution', *Australian Journal of Politics and History*, 29/1, 1983, pp. 50–62.

—— *Trade Unions in China 1949 to the Present: The Organization and Leadership of the All-China Federation of Trade Unions*, Singapore: Singapore University Press, 1986.

LITTLER, CRAIG, R., and LOCKETT, M., 'The Significance of Trade Unions in China', *Industrial Relations Journal*, 14/4, Winter 1983, pp. 31–42.

NG SEK HONG, 'One Brand of Workplace Democracy: a Note on the Workers' Congress in the Chinese Enterprise', *Industrial Relations Journal*, 26/1, Mar. 1984, pp. 56–75.

WALDER, ANDREW G., 'The Remaking of the Chinese Working Class, 1949–1981', *Modern China*, 10/1, Jan. 1984, pp. 3–48.

WARNER, MALCOLM, 'Industrial Relations in the Chinese Factory', *Journal of Industrial Relations*, 29/2, June 1987, pp. 217–32.

WHITE, LYNN T., *Careers in Shanghai*, Berkeley: University of California Press, 1978.

Denmark

GALENSON, WALTER, *The Danish System of Labor Relations*, Cambridge, Mass.: Harvard University Press, 1952.

Finland

KAUPPINEN, TIMO, 'Workplace Cooperation in Finland', *Economic and Industrial Democracy*, 5/1, Feb. 1984, pp. 147–53.

KNOELLINGER, C. E., *Labor in Finland*, Cambridge, Mass.: Harvard University Press, 1960.

RINNE, RAF, 'Industrial Relations in Post-War Finland', *International Labour Review*, 89, May 1964, pp. 461–81.

France

CAIRE, GUY, 'Recent Trends in Collective Bargaining in France', *International Labour Reviews*, 123/6, Nov.–Dec. 1984, pp. 723–42.

LAVAU, GEORGES, 'The Changing Relations between Trade Unions and Working-Class Parties in France', *Goverment and Opposition*, 13, Autumn 1978, pp. 437–57.

REYNAUD, JEAN-DANIEL, 'Trade Unions and Political Parties in France: Some Recent Trends', *Industrial and Labor Relations Review*, 28/2, Jan. 1975, pp. 208–25.

SMITH, W. RAND, 'The Dynamics of Plural Unionism in France: the CGT, CFDT and Industrial Conflict', *British Journal of Industrial Relations*, 22/1, Mar. 1984, pp. 15–33.

India

BAXTER, C., *The Jana Sangh*, Philadelphia: University of Pennsylvania Press, 1969.

BHANDARI, SUNDER SINGH, 'Bharatiya Jana Sang', in L. M. Singhvi (ed.), *Indian Political Parties: Programmes, Promises and Performance*, Delhi: Research, 1971.

BOGAERT, MICHAEL V. D., *Trade Unionism in Indian Ports; A Case Study at Calcutta and Bombay*, New Delhi: Shri Ram Centre for Industrial Relations, 1970.

CHATTERJI, RAKHAHARI, *Unions, Politics and the State: A Study of Indian Labour Politics*, New Delhi: South Asian Publishers, 1980.

CROUCH, HAROLD, *Trade Unions and Politics in India*, Bombay: Manaktalas, 1966.

RAMAN, N. P., *Political Involvement of India's Trade Unions*, London: Asia Publishing, 1967.

RAMASWAMY, E. A., *The Worker and His Union: A Study in South India*, Bombay: Allied Publishers, 1977.

RAMASWAMY, UMA, *Work, Union, and Community: Industrial Man in South India*, Delhi: Oxford University Press, 1983.

ROY, BIREN, 'Charade of Verification', *Economic and Political Weekly* (Bombay), 20–27 Oct. 1984, pp. 1819–20.

SIRSIKA, V. M., and FERNANDES, L., *Indian Political Parties*, New Delhi: Meenakshi, Prakashan, 1984.

VERMA, PRAMOD, and MOOKHERJEE, S., *Trade Unions in India*, New Delhi: Oxford and IBA Publishing, 1982.

Ireland

BOYD, ANDREW, *The Rise of the Irish Trade Unions 1729–1970*, Tralee: Anvil Books, 1972.

CHUBB, BASIL, *The Government and Politics of Ireland*, Stanford: Stanford University Press, 1970.

PRONDZYNSKI, FERDINAND VON, and MCCARTHY, C., *Employment Law*, London: Sweet and Maxwell, 1984.

Israel

BECKER, AHARON, 'The Work of the General Federation of Labour in Israel', *International Labour Review*, 81, May 1960, pp. 436–55.

BEN-PORAT, A., 'Political Parties and Democracy in the Histadrut', *Industrial Relations*, 18/2, Spring 1979, pp. 237–43.

Histadrut: General Federation of Labour in Israel, Tel Aviv: Histadrut, 1984.

INTERNATIONAL LABOUR OFFICE, *Report of the Director-General*, Appendix III ('Report on the Situation of Workers of the Occupied Arab Territories'), Geneva, 1986.

LUTTWAK, EDWARD, and HOROWITZ, D., *The Israeli Army*, London: Allen Lane, 1975.

MEDDING, PETER Y., *Mapai in Israel: Political Organisation and Government in a New Society,* Cambridge: Cambridge University Press, 1972.

PLUNKETT, MARGARET L., 'The Histadrut: the General Federation of Jewish Labor in Israel', *Industrial and Labor Relations Review*, 11/2, Jan. 1958, pp. 155–82.

RESHEF, YONATAN, 'Political Exchange in Israel: Histadrut–State Relations', *Industrial Relations*, 25/3, Fall 1986, pp. 303–19.

ROTHENBERG, GUNTHER E., *The Anatomy of the Israeli Army*, London: Batsford, 1979.

SACHAR, HOWARD M., *A History of Israel*, New York: Knopf, 1976.

SHAARI, YEHUDA, 'Relationship of Histadrut and State', in *The Histadrut*, Supplement (reprint), *Jerusalem Post*, 17 Feb. 1956, pp. 20–3.

ZWEIG, FERDYNAND, 'The Jewish Trade Union Movement in Israel', in S. N. Eisenstadt, R. B. Yosef, and C. Adler (eds.), *Integration and Development in Israel*, Jerusalem: Israel University Press, 1970.

Italy

FARNETI, PAOLO, 'The Troubled Partnership: Trade Unions and Working-Class Parties in Italy, 1948–78', *Government and Opposition*, 13, Autumn 1978, pp. 416–36.

REGINI, MARINO, 'Labour Unions, Industrial Action and Politics', *West European Politics*, 2/3, 1979, pp. 49–66.

WEITZ, PETER R., 'Labor and Politics in a Divided Movement: The Italian Case', *Industrial and Labor Relations Review*, 28/2, Jan. 1975, pp. 226–42.

Japan

COLE, ALAN B., TOTTEN, G. O., UYEHARA, C. H., and DORE, R. P., *Socialist Parties in Post-war Japan*, New Haven: Yale University Press, 1966.

COOK, ALICE H., *An Introduction to Japanese Trade Unionism*, Ithaca: Cornell University Press, 1966.

DORE, RONALD, *British Factory-Japanese Factory: The Origins of National Diversity in Industrial Relations*, Berkeley: University of California Press, 1974.

HANAMI, TADASHI, *Labor Relations in Japan Today*, London: John Martin, 1980.

KAWADA, HISASHI, 'Workers and Their Organizations', in K. Okochi, B. Karsh, and S. B. Levine (eds.), *Workers and Employers in Japan*, Tokyo: University of Tokyo Press, 1984.

MACDOUGALL, TERRY EDWARD (ed.), *Political Leadership in Contemporary Japan*, Ann Arbor: Center for Japanese Studies, University of Michigan, 1982.

ROHLEN, THOMAS P., *For Harmony and Strength: Japanese White-Collar Organization in Anthropological Perspective*, Berkeley: University of California Press, 1974.

STOCKWIN, J. A. A., *Japan: Divided Politics in a Growth Economy*, 2nd edn., London: Weidenfeld and Nicolson, 1982.

WILLEY, RICHARD J., 'Pressure Group Politics: The Case of Sohyo', *Western Political Quarterly*, 17/4, Dec. 1964, pp. 703–23.

Mexico

COLEMAN, KENNETH M., *Diffuse Support in Mexico: The Potential for Crisis*, Beverley Hills: Sage, 1976.

NEEDLER, MARTIN C., *Mexican Politics: the Containment of Conflict*, New York: Praegar, 1982.

ROXBOROUGH, IAN, *Unions and Politics in Mexico: The Case of the Automobile Industry*, Cambridge: Cambridge University Press, 1984.

SCHLAGHECK, JAMES L., *The Political, Economic, and Labor Climate in Mexico*, Philadelphia: Industrial Research Unit, Wharton School, University of Pennsylvania, 1980.

Netherlands

ALBEDA, WIL, 'Changing Industrial Relations in the Netherlands', *Industrial Relations*, 16, May 1977, pp. 133–44.

—— 'Recent Trends in Collective Bargaining in the Netherlands', *International Labour Review*, 124/1, Jan.–Feb. 1985, pp. 49–60.

WINDMULLER, JOHN P., *Labor Relations in the Netherlands*, Ithaca: Cornell University Press, 1969.

New Zealand

GEARE, A. J., *The System of Industrial Relations in New Zealand*, Wellington: Butterworths, 1983.

GUSTAFSON, BARRY, *Social Change and Party Re-organisation: The New Zealand Labour Party since 1945*, London: Sage, 1976.

MILNE, R. S., *Political Parties in New Zealand*, Oxford: Clarendon Press, 1966.

Norway

BALFOUR, CAMPBELL, 'Industrial Relations in Norway', *Industrial Relations Journal*, 5, Summer 1974, pp. 46–53.

INTERNATIONAL LABOUR OFFICE, *The Trade Union Situation and Industrial Relations in Norway*, Geneva, 1984.

MARTIN, PENNY G., 'Strategic Opportunities and Limitations: The Norwegian Labor Party and the Trade Unions', *Industrial and Labor Relations Review*, 28/1, Oct. 1974, pp. 75–88.

Poland

ASCHERSON, NEAL, *The Polish August*, Harmondsworth: Penguin, 1981.

MACSHANE, DENIS, *Solidarity: Poland's Independent Trade Union*, Nottingham: Spokesman, 1981.

STANISZKIS, JADMIGA, *Poland's Self-limiting Revolution* (ed. Jan T. Gross), Princeton: Princeton University Press, 1984.

TOURAINE, ALAIN, DUBET, F., WIEVIORKA, M., and STRZELECKI, J., *Solidarity: The Analysis of a Social Movement, Poland 1980–81*, Cambridge: Cambridge University Press, 1983.

South Africa

DU TOIT, M. A., *South African Trade Unions*, Johannesburg: McGraw-Hill, 1976.

FEIT, EDWARD, *Workers Without Weapons: The South African Congress of Trade Unions and the Organization of the African Workers*, Hamden, Conn.: Archon Books, 1975.

HORRELL, MURIEL, *South Africa's Workers: Their Organizations and the Patterns of Employment*, Johannesburg: SA Institute of Race Relations, 1969.

LEVER, G., 'Trade Unions as a Social Force in South Africa', *Industrial Relations Journal of South Africa*, 1/4, 1981, pp. 36–9.

LODGE, TOM, *Black Politics in South Africa since 1945*, London: Longmans, 1983.

LUCKHARDT, KEN, and WALL, B., *Organise or Starve! A History of the South African Congress of Trade Unions*, London: Lawrence and Wishart, 1980.

MACSHANE, DENIS, PLAUT, M., and WARD, D., *Power! Black Workers, their Unions and the Struggle for Freedom in South Africa*, Nottingham: Spokesman, 1984.

Soviet Union

BROWN, EMILY C., *Soviet Trade Unions and Labor Relations*, Cambridge, Mass.: Harvard University Press, 1966.

CONQUEST, ROBERT, *Industrial Workers in the USSR*, London: Bodley Head, 1967.

DEUTSCHER, ISAAC, *Soviet Trade Unions*, London: Oxford University Press, 1950.

GIDWITZ, BETSY, 'Labor Unrest in the Soviet Union', *Problems of Communism*, 31/6, Nov.–Dec. 1982, pp. 25–42.

HAYNES, V., and SEMYONOVA, O. (eds.), *Workers against the GULAG*, London: Pluto Press, 1979.

KAHAN, ARCADIUS, and RUBLE, B. A. (eds.), *Industrial Labor in the USSR*, Elmsford, NY: Pergamon Press, 1979.

POTICHNYJ, PETER J., *Soviet Agricultural Trade Unions, 1917–70*, Toronto: University of Toronto Press, 1972.

RUBLE, BLAIR A., 'Dual Functioning Trade Unions in the USSR', *British Journal of Industrial Relations*, 17/2, July 1979, pp. 235–41.

—— *Soviet Trade Unions: Their Development in the 1970s*, Cambridge: Cambridge University Press, 1981.

SCHAPIRO, LEONARD, and GODSON, J. (eds.), *The Soviet Worker: From Lenin to Andropov*, London: Macmillan, 1984.

ZIEGLER, CHARLES E., 'Worker Participation and Worker Discontent in the Soviet Union', *Political Science Quarterly*, 98/2, Summer 1983, pp. 235–53.

Sweden

CARLSON, BO, *Trade Unions in Sweden*, Stockholm: Tidens forlag, 1969.

ELVANDER, NILS, 'In Search of New Relationships: Parties, Unions, and Salaried Employees' Associations in Sweden', *Industrial and Labor Relations Review*, 28/1, Oct. 1974, pp. 60–74.

GUSTAFSON, AGNE, 'Rise and Decline of Nations: Sweden', *Scandinavian Political Studies*, 9/1, Mar. 1986, pp. 35–50.

HUNTFORD, ROLAND, *The New Totalitarians*, New York: Stein and Day, 1972.

KORPI, WALTER, *The Working Class in Welfare Capitalism: Work, Unions, and Politics in Sweden*, London: Routledge, 1978.

LASH, SCOTT, 'The End of Neo-corporatism?: the Breakdown of Centralised Bargaining in Sweden', *British Journal of Industrial Relations*, 23/2, July 1985, pp. 215–39.

Tanzania

DONGE, JAN KEES VAN, and LIVIGA, A. J., 'In Defence of the Tanzanian Parliament', *Parliamentary Affairs*, 39/2, Apr. 1986, pp. 230–40.

FRIEDLAND, WILLIAM H., *Vuta Kamba: The Development of Trade Unions in Tanganyika*, Stanford: Hoover Institution Press, 1969.

KARIOKI, JAMES N., *Tanzania's Human Revolution*, University Park: Pennsylvania State University Press, 1979.

NYERERE, JULIUS K., *Freedom and Development: A Selection from Writings and Speeches 1968–1973*, Dar Es Salaam: Oxford University Press, 1973.

SAMOFF, JOEL, *Tanzania: Local Politics and the Structure of Power*, Madison: University of Wisconsin Press, 1974.

TUMBO, N. S. K., *et al.*, *Labour in Tanzania*, Dar Es Salaam: Tanzania Publishing House, 1977.

YEAGER, RODGER, *Tanzania: An African Experiment*, Boulder, Colorado: Westview Press, 1982.

United Kingdom

ALLEN, V. L., *Trade Unions and the Government*, London: Longmans, 1960.

BARRIES, DENIS, and REID, E., *Governments and Trade Unions: The British Experience 1964–79*, London: Heinemann, 1980.

BEER, SAMUEL H., *Britain Against Itself: The Political Contradictions of Collectivism*, London: Faber, 1982.

CLEGG, H. A., *The Changing System of Industrial Relations in Great Britain*, Oxford: Blackwell, 1979.

COATES, KEN, and TOPHAM, T., *Trade Unions and Politics*, Oxford: Blackwell, 1986.

—— *Trade Unions in Britain*, Nottingham: Spokesman, 1980.

CROUCH, COLIN, 'The Peculiar Relationship: The Party and the Unions', in D. Kavanagh (ed.), *The Politics of the Labour Party*, London: Allen and Unwin, 1982.

—— *The Politics of Industrial Relations*, 2nd edn., London: Fontana, 1982.

HARRISON, MARTIN, *Trade Unions and the Labour Party since 1945*, London: Allen and Unwin, 1960.

HOUGHTON, DOUGLAS, 'Trade Union MPs in the British House of Commons', *The Parliamentarian*, Oct. 1968, pp. 215–21.

KIMBER, RICHARD, and RICHARDSON, J. J. (eds.), *Pressure Groups in Britain*, London: Dent, 1974.

LEWIS, ROY (ed.), *Labour Law in Britain*, Oxford: Blackwell, 1986.

MAY, TIMOTHY C., *Trade Unions and Pressure Group Politics*, Westmead: Saxon House, 1975.

MINKIN, LEWIS, *The Labour Party Conference*, London: Allen Lane, 1978.

—— 'The Party Connection: Divergence and Convergence in the British Labour Movement', *Government and Opposition*, 13, Autumn 1978, pp. 458–84.

TURNER, H. A., *Trade Union Growth, Structure and Policy: A Comparative Study of the Cotton Unions in England*, London: Allen and Unwin, 1962.

United States

AFL–CIO, *The Changing Situation of Workers and Their Unions*. Report by the AFL–CIO Committee on the Evolution of Work, Feb. 1985.

BOK, DEREK C., and DUNLOP, JOHN T., *Labor and the American Community*, New York: Simon and Schuster, 1970.

FERMAN, LOUIS A. (ed.), 'The Future of American Unionism', in *The Annals of the American Academy of Political and Social Science*, 473, May 1984, pp. 9–189.

GALENSON, WALTER, *The CIO Challenge to the AFL: A History of the American Labor Movement 1935-1941*, Cambridge, Mass.: Harvard University Press, 1960.

GOLDFIELD, MICHAEL, 'Labour in American Politics—its Current Weakness', *Journal of Politics*, 48/1, Feb. 1986, pp. 2–29.

GOLDMAN, ALVIN L., *Labor Law and Industrial Relations in the United States of America*, 2nd edn., Antwerp: Kluwer, 1984.

GREENSTONE, J. DAVID, *Labor in American Politics*, New York: Knopf, 1969.

HOXIE, ROBERT F., *Trade Unionism in the United States*, New York: Russell, 1966.

MAHOOD, H. R., *Pressure Groups in America*, New York: Charles Scribener's Sons, 1967.

MASTERS, NICHOLAS A., 'The Organized Labor Bureaucracy as a Base of Support for the Democratic Party', *Law and Contemporary Problems*, 27, Spring 1962, pp. 252–65.

MILLER, RONALD, 'The Mid-Life Crisis of the American Labor Movement', *Industrial Relations Journal*, 18/3, Autumn 1987, pp. 159–69.

PERLMAN, SELIG, *A History of Trade Unionism in the United States*, New York: Kelley, 1950.

RADOSH, RONALD, *American Labor and United States Foreign Policy*, New York: Random House, 1969.

REHMUS, CHARLES M., MCLAUGHLIN, DORIS B., and NESBITT, FREDERICK H.. (eds.), *Labor and American Politics: A Book of Readings*, Ann Arbor: University of Michigan Press, 1978.

TAWNEY, R. H., *The American Labour Movement and Other Essays* (ed. J. M. Winter), Brighton: Harvester Press, 1979.

VALE, VIVIAN, *Labour in American Politics*, New York: Barnes and Noble, 1971.

WILSON, GRAHAM K., *Unions in American Politics*, London: Macmillan, 1979.

WINDMULLER, JOHN P., 'Labor: A Partner in American Foreign Policy?', *The Annals of the American Academy of Political and Social Science*, 350, Nov. 1963, pp. 104–14.

ZEIGLER, HARMON, *Interest Groups in American Society*, Englewood Cliffs: Prentice-Hall, 1964.

Venezuela

BLANK, DAVID E., *Politics in Venezuela*, Boston: Little, Brown, 1973.

HERMAN, DONALD L., *Christian Democracy in Venezuela*, Chapel Hill: University of North Carolina Press, 1980.

MARTZ, JOHN D., *Acción Democratica: Evolution of a Modern Political Party in Venezuela*, Princeton: Princeton University Press, 1966.

VALENTE, CECILIA M., *The Political, Economic, and Labor Climate in Venezuela*, Philadelphia: Industrial Research Unit, Wharton School, University of Pennsylvania, 1979.

YEPES, JOSE ANTONIO GIL, *The Challenge of Venezuelan Democracy*, New Brunswick: Transaction Books, 1981.

West Germany

BEYME, KLAUS VON, 'The Changing Relations between Trade Unions and the Social Democratic Party in West Germany', *Government and Opposition*, 13/4, Autumn 1978, pp. 399–415.

CULLINGFORD, E. C. M., *Trade Unions in West Germany*, Boulder, Colorado: Westview Press, 1976.

ESSER, JOSEF, 'State, Business and Trade Unions in West Germany after the "Political Wende" ', *West European Politics*, 9/2, Apr. 1986, pp. 198–214.

MARKOWITS, ANDREI S., *The Politics of the West German Trade Unions*, Cambridge: Cambridge University Press, 1986.

WILLEY, RICHARD J., 'Trade Unions and Political Parties in the Federal Republic of Germany', *Industrial and Labor Relations Review*, 28/1, Oct. 1974, pp. 38–59.

Yugoslavia

CARTER, APRIL, *Democratic Reform in Yugoslavia: The Changing Role of the Party*, London: Frances Pinter, 1982.

COMISSO, ELLEN, *Workers' Control under Plan and Market*, New Haven: Yale University Press, 1979.

—— 'Workers' Councils and Labor Unions: Some Objective Trade Offs', *Politics and Society*, 10/3, 1981, pp. 251–79.

JOVANOV, NECA, 'Strikes and Self-management', in J. Obradovic and W. N. Dunn (eds.), *Workers' Self-Management and Organizational Power in Yugoslavia*, Pittsburgh: Center for International Studies, University of Pittsburgh, 1978.

KOLAJA, JIRI, *Workers' Councils: The Yugoslav Experience*, London: Tavistock, 1965.

LYDALL, HAROLD, *Yugoslav Socialism: Theory and Practice*, Oxford: Clarendon Press, 1984.

SINGLETON, FRED, *Twentieth-century Yugoslavia*, London: Macmillan, 1976.

ZUKIN, SHARON, 'The Representation of Working-Class Interests in Socialist Society: Yugoslav Labor Unions', *Politics and Society*, 10/3, 1981, pp. 281–316.

Other

ANDRADE, LUIS F., *The Labor Climate in Brazil*, Philadelphia: Industrial Research Unit, Wharton School, University of Pennsylvania, 1985.

ANGELL, ALAN, *Politics and the Labor Movement in Chile*, London: Oxford University Press, 1972.

AYUBI, NAZIH N. M., *Bureaucracy and Politics in Contemporary Egypt*, London: Ithaca Press, 1980.

DUNKLEY, GRAHAM, 'Industrial Relations and Labour in Malaysia', *Journal of Industrial Relations*, 24/3, Sept. 1982, pp. 424–42.

FISHMAN, ROBERT M., 'The Labor Movement in Spain: From Authoritarianism to Democracy', *Comparative Politics*, 14/3, Apr. 1982, pp. 281–305.

GDR: 300 Questions 300 Answers, Dresden: German Institute of Contemporary History, n.d.

Hince, K. W., 'Trade Unionism in Fiji', *Journal of Industrial Relations*, 13, Dec. 1971, pp. 368–89.

Jacobs, J. Bruce, 'Political Opposition and Taiwan's Political Future', *Australian Journal of Chinese Affairs*, 6, July 1981, pp. 21–44.

Jecchinis, Christos, *Trade Unionism in Greece: A Study in Political Paternalism*, Chicago: Labor Education, Roosevelt University, 1967.

Johnson, Nancy R., *The Political, Economic, and Labor Climate in Peru*, Philadelphia: Industrial Research Unit, Wharton School, University of Pennsylvania, 1982.

Kearney, Robert N., *Trade Unions and Politics in Ceylon*, Berkeley: University of California Press, 1971.

Kim, Kyong-Dong, 'Socio-economic Changes and Political Selectivity in the Development of Industrial Democracy in the Republic of Korea', *Economic and Industrial Democracy*, 5/4, Nov. 1984, pp. 445–67.

Landau, Jacob M., *Radical Politics in Modern Turkey*, Leiden: E. J. Brill, 1974.

Lomax, Bill, *Hungary 1956*, London: Allison and Busby, 1976.

Magill, John H., *Labor Unions and Political Socialization: A Case Study of Bolivian Workers*, New York: Praegar, 1974.

Reeve, David, *Golkar of Indonesia*, Singapore: Oxford University Press, 1985.

Schlagheck, James L., *The Political, Economic, and Labor Climate in Brazil*, Philadelphia: Industrial Research Unit, Wharton School, University of Pennsylvania, 1977.

Siegenthaler, Jurg K., 'Current Problems of Trade Union–Party Relations in Switzerland: Re-orientation versus Inertia', *Industrial and Labor Relations Review*, 28/2, Jan. 1975, pp. 265–81.

Tzannatos, Zafiris, *Socialism in Greece: The First Four Years*, Aldershot: Gower, 1986.

Vickery, Michael, *Kampuchea: Politics, Economics and Society*, London: Frances Pinter, 1986.

Windmuller, John P., 'Czechoslovakia and the Communist Union Model', *British Journal of Industrial Relations*, 9/1, Mar. 1971, pp. 33–54.

Wynia, Gary W., *Argentina: Illusions and Realities*, New York: Holmes and Meier, 1986.

Index

The titles of national trade union confederations, political parties, government bodies, and legislation are all indexed by country.